The Beautiful Cure

ALSO BY DANIEL M. DAVIS

The Compatibility Gene

The Beautiful Cure

HARNESSING YOUR BODY'S NATURAL DEFENCES

Daniel M. Davis

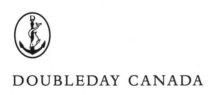

DOUBLEDAY CANADA

Doubleday Canada and colophon are registered trademarks of Penguin Random House Canada Limited

Published by arrangement with Bodley Head, one of the publishers in the Random House Group Ltd.

Library and Archives Canada Cataloguing in Publication

Davis, Daniel M. (Daniel Michael), 1970-, author
The beautiful cure / Daniel M. Davis.

Issued in print and electronic formats.
ISBN 978-0-385-68676-1 (hardcover).--ISBN 978-0-385-68617-4 (EPUB)

1. Immune system--Popular works. 2. Medicine, Preventive--Popular works. I. Title.

QR181.7.D35 2018   616.07'9   C2017-906877-6
                                                         C2017-906878-4

Jacket design by Five Seventeen
Jacket images: © Getty Images
Typeset in India by Integra Software Services Pvt. Ltd, Pondicherry

Printed and bound in the USA

Published in Canada by Doubleday Canada,
a division of Penguin Random House Canada Limited

www.penguinrandomhouse.ca

10 9 8 7 6 5 4 3 2 1

Penguin
Random House
DOUBLEDAY CANADA

In memory of Jack and Ruby Faulkner

Contents

A note to professional scientists xi
Overview 1

PART ONE: The Scientific Revolution in Immunity

1 Dirty Little Secrets 9
2 The Alarm Cell 32
3 Restraint and Control 57
4 A Multibillion-Dollar Blockbuster 83

PART TWO: The Galaxy Within

5 Fever, Stress and the Power of the Mind 109
6 Time and Space 129
7 The Guardian Cells 149
8 Future Medicines 172

Epilogue 196
Acknowledgements 198
Notes 200
Index 253

There are mysteries which men can only guess at, which age by age they may solve only in part. Believe me, we are now on the verge of one.

Bram Stoker, Dracula (1897)

A note to professional scientists

Immunology is an extraordinarily rich subject and I can only apologise to any scientist whose contributions I have not included or have mentioned all too briefly. As P. G. Wodehouse wrote in *Summer Moonshine* (1937): 'It is one of the inevitable drawbacks to a narrative like this one that the chronicler, in order to follow the fortunes of certain individuals, is compelled to concentrate his attention on them and so to neglect others equally deserving of notice.' Through interviews with the scientists involved and my own reading of the original research I have sought to describe how advancements were made, but any one book can only ever tell part of a story.

Overview

'Look at that flower; look how beautiful it is,' said an artist to his friend. 'Art appreciates and celebrates that beauty, whereas science just takes it all apart. Science makes the flower dull.'

The friend being addressed was Nobel Prize-winning physicist Richard Feynman, and he thought that the artist's view was 'a bit nutty'. Feynman countered that he too could appreciate the flower's beauty, but as a scientist he knew that the inner structure of the flower is wondrous as well – with its cells, its chemical and biological processes, all of its many intricate systems. In addition, Feynman explained, knowing that the flower attracts insects we might deduce that insects find the flower aesthetically pleasing, which in turn raises all sorts of questions about evolution, cognition and light. 'Science,' Feynman said, 'only adds to the excitement and mystery and the awe of a flower. It only adds.'[1]

Feynman related this now-famous exchange in an interview on BBC TV in 1981, when I was aged eleven. I already knew that I wanted to be a scientist, but Feynman, with his strong New York accent and with roses swaying in the window behind him, captured the reason better than I could say myself. Now, leading a team of researchers to study human immune cells in minute detail, I've seen first-hand how science reveals beauty where otherwise it might have remained hidden. The inside of the human body may not have evolved to be aesthetically pleasing like a flower, but splendour ascends from its details.

In all of human biology, the process that's been studied the most, details excavated the deepest, is the body's response to a cut or an infection. The familiarity of the symptoms – redness, tenderness and inflammation – belies wonders taking place beneath the skin, where swarms of different cells move in to fight off germs, as well as repair the damage and deal with the debris. Far from conscious control, this reflex is essential for our survival.

A simple view of what is happening here is that the body attacks germs which invade the wound because our immune system is programmed to fight whatever is not part of us. But a moment's reflection shows that this cannot be the whole story. Food isn't part of your body and yet your immune system mustn't react to everything you eat. More subtly, your immune system must be able to tell the difference between friendly bacteria that live in your gut, which should be left alone, and dangerous bacteria that might make you sick, which need to be dealt with.

This crucial realisation, that an immune response can't be triggered by just anything alien to the human body, came only as recently as 1989, and it would take many more years before a deeper understanding emerged. In the meantime, a pains-taking, game-changing scientific adventure unfolded in which the world of immunity has opened up to reveal what it really is: not a simple circuit involving a few types of immune cells but a multilayered, dynamic lattice of interlocking subsystems, one of the most complex and important frontiers of scientific enquiry we know of. As this book will show, the many discoveries that have resulted from this adventure amount to a scientific revolution in our understanding of the human body and are set to spark a revolution for medicine in the twenty-first century.

For a start, we have come to realise that our body's ability to fight disease is continuously changing. The power of our immune system waxes and wanes, affected by stress, old age, the time of day and our state of mind. Our immune system is in constant flux; our health balanced on a tightrope. For example, the number of immune cells in our blood tends to peak in the evening and is at its lowest in the morning. There are many changes that happen to our immune system during the night, as our body enters a different state of activity and energy use, and in turn our immune system seems to be affected by how well we sleep. Reduced sleep – less than five hours per night – correlates with an increased risk of the common cold and pneumonia.[2] Among other things, this book will explore the effects of night-shift work on our immune system and whether or not practices that could reduce stress, such as t'ai chi or mindfulness, are able to help us fight infections.

Mysteries remain but already these discoveries challenge the simple view we once held about how our bodies fight disease – and what it takes to be healthy. Even though it's correct – very roughly – that the immune system targets what's not part of you, it has become apparent that layer upon layer of biological checks and balances, run by countless cells and molecules, regulate the process. Resolving the mysteries and the complexities allows us to approach questions of major importance to our health and well-being: why do some people get cancer and can our immune system fight it? How do vaccines work and can we make them better? What exactly is an autoimmune disease and what can we do about it? The vast majority of ailments that afflict us are cured by our bodies' natural defences. Understanding and harnessing this power might turn out to be the one of the most important gifts that science gives for the health of humankind.

While some drugs, such as penicillin, kill germs directly, many human maladies, from cancer to diabetes, may be fought best with new kinds of medicine that enhance (or in some cases suppress) the activity of our immune system. Unlike penicillin, and medicines like it that are made naturally – by a fungus in the case of penicillin – and merely *isolated* by scientists, these new medicines that work with our immune system are *designed* by scientists. Scientists studying the immune system can have ideas which become therapies and multibillion-dollar drugs. But these medicines must be tuned to work with utmost precision. If we over-activate the immune system, healthy cells and tissues will be destroyed, and if we switch it off entirely we will become susceptible to all kinds of germs that are normally easily dealt with. The potential benefits are transformative but the consequences when things go wrong can be terrible.

The vast endeavour to understand immunity has also thrown up new insights into many other areas of human biology too, such as the ageing process. 80–90% of people who die because of the flu virus are over sixty-five.[3] Why is it that as we age, our defence against infections grow weaker? Why do we heal less easily and are more likely to succumb to autoimmune diseases? We have learnt that part of the problem is that the elderly have fewer of some types of immune cells circulating in their blood. Another is that immune cells in the elderly are worse at detecting disease. Compounding the challenges of ageing itself is the fact that the elderly often struggle with sleep-deprivation and stress, which also affect our immune system. Working out how much each of these various factors affects our health can be extremely difficult because it is almost impossible to isolate any one of them. While stress affects our immune system, it also correlates with sleep loss, making it hard to know the effect of either on its own.

In fact, pretty much everything in the body is connected with everything else – even more so than you might imagine. It has recently emerged that the immune system is intimately connected with a huge range of diseases that appear to be unrelated to its role in fighting germs: heart problems, neurological disorders, even obesity. My first book, *The Compatibility Gene*, discussed one element of the immune system, a handful of genes that influence our individual responses to infections. *The Beautiful Cure* is about the bigger picture: how and why the activity of our immune system varies, how it is regulated and directed, all of its component parts – the whole shebang.

This is also a book about how scientific ideas develop. The quest to understand immunity is one of humankind's greatest scientific adventures, and the impersonal knowledge we have now has been won through a saga of personal hardships, triumphs and sacrifices. Many men and women have devoted their careers, and much of their lives, to understanding a mere fragment of the whole. This quest has created many deep friendships; passion for science can be a powerful bond. On the other hand, there are a few scientists involved who now can't stand to be in same room together. Countless researchers have contributed, each making wondrous discoveries about particular cells or molecules in our immune system, but in the end, any one person's contribution is small – even the geniuses' – and the sacrifices some scientists have made might seem out of all proportion, beyond what most people would be willing to accept.

My own research involves using specialist microscopes to watch what happens at the point of contact between immune cells when they interact with one another, and to watch the contacts immune cells make with other cells to decide whether they are healthy or diseased. My discoveries have helped show how immune cells communicate with each other and how they detect signs of disease in other cells, which in turn helps us understand precisely how the immune system is regulated. We each add a little, focusing on one part of the system at a time.

When we divide an integrated system into separate elements in this way, it doesn't make it dull – as Richard Feynman's artist friend thought – but it's not entirely fulfilling either. Things work together and each component only makes sense when it's seen as part of a whole. Textbooks about the immune system tend to discuss the role of each molecule or cell in turn, but that's like explaining a bicycle by describing what a wheel is, and then what a handlebar is, and then what a brake is. None of these single elements are properly understood without the others; their meaning lies in the relationships between them. Just as much as the parts build up a system, the system defines the parts. We marvel at the details but we must also pan out to the big picture, because it's only when we do this that we can begin to exploit our knowledge of immunity for a revolution in health.

We will explore that revolution in the second half of the book. First, *The Beautiful Cure* charts the global scientific adventure that has led to it, revealing a world of unsung heroes and rebels who have discovered how and why the immune system works the way it does. If solace or joy can ever be gleaned from the beauty of nature, then what they have uncovered – the complexity, delicacy and elegance of our immune system – is as inspirational as any other frontier of science, from the substructure of atoms to the birth of stars.

Part One

The Scientific Revolution in Immunity

1 Dirty Little Secrets

What does it take to do something great? In 2008, an experiment was conducted in which experienced chess players were shown a game that could be won using a well-known sequence of five moves. But there was also a more dramatic, unconventional way to win the same game, in just three moves. When asked for the quickest way to win the game, experts usually pointed out the familiar five-move plan, missing the optimal three. Only the very best chess players – the grandmasters – saw the three-move win; ordinary experts stuck to what they were familiar with.[1]

It's in our nature to try to resolve problems using what has worked before. But knowing what worked before can blind us to the insight required for major leaps forward.[2] Our greatest scientists are those who, despite their expertise, remain free to think differently. By this measure, Charles Janeway, an immunologist working at Yale University, was indeed one of our greatest scientists. He was also said to be 'one of the most exciting, decent, and thoughtful immunologists on the planet'.[3]

Born in Boston in 1943, Janeway studied chemistry and then medicine at Harvard. His path to medicine was influenced by his father, an eminent Harvard paediatrician and a department head at Boston's Children's Hospital,[4] but Janeway felt that 'surgery was going to condemn [him] to a life of routine procedures'[5] and he switched his focus to basic research. He married when young but in 1970, aged twenty-seven, he split up from his wife Sally, when their child was aged one. As a result, he 'felt lonely for many years',[6] but gained time and freedom for his research. In 1977 he joined the faculty of Yale, where he met his second wife, Kim Bottomly, also a well-known immunologist.

In 1989, Janeway puzzled over what he called the 'dirty little secret' in our understanding of immunity. The problem concerned vaccines and the way in which they were thought to work. The basic principle of vaccination follows the familiar idea that an infection, caused by a virus or bacteria, is dealt with much more efficiently if your immune system has encountered that same virus or bacteria previously. So – the dogma goes – vaccines work by exposing you to a dead or harmless version of a germ. By provoking your immune system to build up defences against it, it prepares you to respond rapidly if you encounter the same germ again. This works because the particular immune cells that are activated by a particular germ multiply and persist in the body for a long time, long after the germ has been eradicated, meaning they are ready for action if they encounter the same germ again. And with this, so it seems, one of humankind's greatest medical triumphs can be explained in just a few lines.

But take one step deeper and it turns out that vaccination has a touch of alchemy about it too. The 'dirty little secret' is that vaccines only work well when so-called 'adjuvants' are added. Adjuvants (from the Latin word *adiuvare* meaning 'to help') are chemicals, such as aluminium hydroxide, which, as discovered by chance, help vaccines be effective. At one level it seems such a small thing – aluminium hydroxide somehow helps vaccines be effective – but to Janeway this small technical tip revealed a crack in our basic understanding, because no one could actually explain *why* adjuvants did this. Understanding vaccination is unquestionably important – nothing apart from providing safe water, not even antibiotics, has ever saved more lives[7] – and Janeway was determined to understand precisely why adjuvant was necessary. In doing so, he uncovered a whole new way of thinking about how the human immune system *really* works.

★

The use of vaccination as a medical procedure long pre-dates any scientific knowledge of how the process works. The first descriptions of this vital life-saver can be found in folklore.[8] Deliberate infections to provide protection – inoculations – were practised in China, India and some African countries, long before any formal medical procedure was established.[9] The scientific story begins, however, in 1721, when an epidemic of smallpox made the British royal family anxious, especially for the safety of their children. The royals had heard of rural traditions and stories from other countries about how to inoculate against the disease, but details varied as to how, exactly, the procedure should be performed. Was an application of blister fluid best? Or were hand-squeezed smallpox scabs preferable? It was widely known that people only ever got smallpox once, and so the real issue was whether or not a small dose of smallpox could be given to someone without it killing them. A test was needed to determine the safety and efficacy of inoculation before it was used on the royal family – and prisoners seemed appropriate for the honour.

The first recorded 'clinical trial' in the history of immunity[10] was performed on 'volunteers' who had been recruited on the basis that they could either participate in the potentially deadly trial or face the certain death of judicial execution. On 9 August 1721, incisions were made on the arms and legs of six convicts. Skin and pus from a smallpox patient was rubbed in. Another prisoner was given a sample of skin and pus up her nose – needless to say, to her great discomfort. Twenty-five members of the scientific elite witnessed the event, including fellows of the Royal Society (which had been granted its Royal Charter in 1662 but still only had vague criteria for membership).[11] In accordance with folk wisdom, each prisoner became ill with symptoms of smallpox for a day or two, and then recovered. The woman inoculated nasally became especially ill, but recovered nonetheless.[12] On 6 September 1721, King George I pardoned the convicted volunteers and they were released. Their immune systems had saved them from two death sentences: the gallows and the pox.

A few months later on 17 April 1722, the Prince and Princess of Wales – who in five years would become King George II and Queen Caroline – had two of their own daughters inoculated.[13] The event was covered by all the newspapers and led to considerable interest in inoculation (a reminder that high-profile leaders or celebrities have enormous influence on public attitudes to new scientific ideas).[14] Even so, the procedure remained controversial, in part because, some claimed, the intervention went against Nature or God – a London preacher spoke in 1722 on 'the dangerous and sinful practice of inoculation' for example – but also because around 2% of people died after being deliberately inoculated with smallpox.[15]

Forty-eight years later, a twenty-one-year-old man named Edward Jenner began three years' training at St George's Hospital, London, under John Hunter, one of the most prominent surgeons and anatomists in England. Hunter helped sharpen Jenner's critical faculties and cultivated his passion for experimentation, but he never got to see how his protégé blossomed. Hunter died in 1793, three years before Jenner discovered a way to circumvent the acute danger of inoculation while achieving the same effect.

As a country physician who spent most of his life in his small hometown of Berkeley, Gloucestershire, Jenner was familiar with the fact that milkmaids never get smallpox. His revelatory idea was that perhaps their exposure to cowpox – a mild viral infection that humans could catch from cows – provided protection against smallpox, and that pus from non-fatal cowpox blisters might therefore be used for inoculation instead of pus from smallpox victims, which was far more dangerous. His now-legendary experiment was performed on 14 May 1796. Jenner took pus from dairymaid Sarah Nelmes, who had been infected with cow pox from one of her cows, Blossom, and inoculated his gardener's eight-year-old son, James Phipps. James was then given pus from a smallpox patient and he didn't get ill.

This experiment is often said to mark the birth of immunology but at the time, Jenner had trouble just publishing his findings. The Royal Society said the observation was merely anecdotal – which it was – and suggested that Jenner should first test many more children before making such bold claims. Jenner did repeat the test on others, including his own eleven-month-old son, but even so, he didn't try to publish with the Royal Society again. Instead, Jenner self-published his work in a large-print seventy-five-page book. Initially available in just two London shops, the book was released on 17 September 1798 and became a huge success.[16] The term 'vaccine' was coined a few years later by a friend of Jenner to describe the process he had discovered, from the Latin word for cow, *vacca*.[17] Smallpox became the first disease fought on a global scale and was officially eradicated in 1980.[18]

Jenner always believed that his work could lead to a global annihilation of smallpox, but he never had a deep understanding of how vaccination worked.[19] By the time of Janeway's epiphany in 1989, the consensus view was that the presence of a germ in the body triggers an immune reaction because the body is primed to detect molecules that it has not encountered before; in other words, that the immune system works by reacting against molecules that are *non-self* – not from the body.[20] After exposure to molecules alien to the body, the immune system is poised to react rapidly if the same non-self molecules are encountered again. But an experiment performed by two different scientists working independently in the early 1920s (it's unclear precisely when),[21] didn't fit this simple view of vaccination, and this puzzled Janeway deeply.

The experiment was performed by French biologist Gaston Ramon and London physician Alexander Glenny. They each discovered that a protein molecule made by the bacteria which cause diphtheria – diphtheria toxin – could be inactivated by heat and a small amount of the chemical formalin. Potentially this meant it might be used as a safe vaccine against the disease. To their surprise, however, when the inactivated protein molecule was injected into animals, the immunity it produced was only short-lived. The observation was seen as mildly curious at the time, and largely forgotten, but decades later Janeway reasoned that protein from the bacteria was non-self – not part of the human body – and so, according to the consensus view of the 1980s, there was no explanation for why it would not work well as a vaccine. How come the pus from cowpox blisters worked well as a vaccine, Janeway wondered, whereas protein molecules such as diphtheria toxin, which had been isolated from germs, did not?

Glenny was a workaholic, and though extremely shy and not easy to get on with, he was skilled in organising his research, streamlining procedures so that he and his co-workers could perform huge numbers of experiments with great efficiency.[22] He had no time for proper statistical analysis; results were either 'obvious and useful, or doubtful and valueless'.[23] This attitude – go-getting, fast-moving – was an important factor in his lab's ability to screen an enormous number of experimental conditions, seeking a way to make diphtheria toxin work as a vaccine.[24] Eventually, in 1926, Glenny's team found that when diphtheria protein was purified by a chemical process that involved combining it with aluminium salts, it became an effective vaccine. Glenny's explanation was that the aluminium salts helped the diphtheria toxin stay in the body long enough for an immune reaction to develop, but no one knew of any process which could explain how or why this might be.[25] After Glenny, other substances such as paraffin oil were discovered to help vaccines work in the same way that aluminium salts did, and collectively they became known as adjuvants. But still, there was no obvious common feature that explained *why* they worked.

In January 1989, Janeway and his wife, fellow immunologist Kim Bottomly, were discussing what happens in the body when you get a cut or an infection. They realised that they could not easily explain how an immune response starts: what exactly was the trigger? As Bottomly recalled, they often argued about scientific matters in their car and later simply forgot what had been said, but this time they were attending a conference in Steamboat Springs, Colorado, so they had their notebooks with them.[26] The debate stuck with Janeway. For the next few months, he mulled over the problem – how does an immune reaction start? – as well as the question of how adjuvants work, and it was by thinking about the two problems together that he had a revelatory idea.

An important clue was that a chemical normally found in the outer coating of bacteria (a large molecule with the cumbersome name of lipopolysaccharide, or LPS) had been shown to be an especially effective adjuvant. What if, Janeway reasoned, the presence of something that has never been in your body before was not the *sole* indication that an immune reaction should occur? What if there has to be something else – *a second signal* – that's needed to kick off an immune reaction, a second signal that can be provided by an adjuvant, which might in turn replicate the presence of actual germs? This might explain why protein molecules separated from their originating germ were ineffective as vaccines, but a molecule such as LPS, from the outer coating of bacteria, worked well as an adjuvant.

With great gusto, Janeway first presented his idea in a now-famous paper entitled 'Approaching the asymptote? Evolution and revolution in immunology', published in the proceedings of a prestigious meeting at Cold Spring Harbor, New York, held in June 1989.[27] In it he suggested that everyone seemed to be studying the immune system as if the knowledge was approaching 'some sort of asymptote, where future experiments are obvious, technically difficult to perform, and aim to achieve ever higher degrees of precision rather than revolutionary changes in our understanding'.[28] As a result, they had all missed something big: the 'tremendous gap' in our understanding of how immune reactions start.[29] He suggested that distinguishing between self and non-self was not enough: the immune system has to be able to tell when something is likely to be a threat to the body before an immune reaction takes place, and that therefore the immune system must, he reasoned, be able to detect telltale signs of actual germs or infected cells. He predicted that there had to be a whole part of our immune system, yet to be identified, with this very purpose, and he even predicted a way it could work.

As we have seen and as Janeway pointed out, nobody at this time paid much attention to how an immune reaction started, and most (if not all) researchers focused on understanding another aspect of immunity, related to inoculation and vaccination: namely, how the immune system is able to respond to germs faster and more efficiently a second time around. It was known that at the heart of this process are two types of white blood cells called T cells and B cells. These white blood cells have an especially important receptor molecule at their surface, not so imaginatively called the T cell receptor and the B cell receptor. These receptors come from the class of biological molecules known as proteins, which are long strings of atoms that fold up into elaborate shapes well adapted for a specific task in the body. In general, proteins bind or join with other molecules, including other proteins, to complete their tasks, and the precise shape of a protein dictates which types of other molecules it is able to connect with, in the same way that two jigsaw pieces interlock by having complementary shapes. The receptor on each individual T cell or B cell has a slightly different shape, allowing it to interlock with a different foreign molecule. It reaches out from the immune cell's surface into its surroundings, and if it connects with something that hasn't been in your body before, it 'switches on' the immune cell, which then kills the germ or infected cell directly, or summons other immune cells to help. Crucially, the activated immune cell also multiplies, populating your body with more cells that have the same usefully shaped receptor. Some of these cells stay in the body for a long time, which is what gives the immune system a *memory* for germs that have been encountered before – which is, of course, at the heart of how vaccination works.

Importantly, receptors on T cells and B cells are not made to bind germs per se; these receptors have randomly shaped ends, which allow them to lock onto all kinds of molecules. The way in which the body ensures they only latch onto germs is one of the greatest wonders of the immune system, and works as follows. Each T cell and B cell acquires its receptor while developing in bone marrow. A shuffling of genes as the cell develops gives each cell a uniquely shaped receptor. But before entering the bloodstream, each individual T cell and B cell is tested in case its receptor is able to bind to healthy cells. If it is, then that particular T cell or B cell is killed off, because it would be dangerous to have such an immune cell in the body. In this way, only T cells and B cells that won't attack healthy cells are allowed to defend the body, and by the same logic, if a receptor on a T cell or B cell does bind to something, that something must be a molecule that hasn't been in your body before. In formal language, this is how the immune system is able to distinguish *self*, the components of your body, from *non-self*, anything that's not part of you.

What Janeway predicted was that this is not the whole story. Specifically, he predicted that there must be receptors (which he called *pattern-recognition receptors*) that are not randomly generated and then selected, but rather have fixed shapes that interlock specifically with germs or infected cells (or rather with *molecular patterns* that are found only on germs or infected cells).[30] Because this appeared a much simpler way for immune cells to detect germs compared with the elaborate process of making immune cells with randomly shaped receptors and then killing off those which might react against healthy cells, Janeway suggested that receptors with fixed shapes probably evolved first to defend against disease, and only later, when life on earth became more complex, did a more elaborate immune system develop, which then included T cells and B cells.

The simpler system of fixed pattern-recognition receptors that Janeway predicted forms part of the system often called *innate* immunity, by contrast with the aspect of our immune defence that accounts for its memory of past infections, which is called *adaptive* immunity. The term 'innate immunity' was already used before Janeway – to describe early-acting defence mechanisms provided by skin, mucus and the immediate actions of immune cells that move into a cut or wound – but the subject was given only a few pages in textbooks, including the bestselling one written by Janeway himself.[31] What made Janeway's ideas revolutionary was that he essentially changed the immune system's mission statement. Before Janeway, the *raison d'être* of the immune system was to react against things that have never been in your body before. But Janeway said that the immune system must respond to things that haven't been in your body before – *and are from germs.*

In hindsight, it's blatantly necessary that the immune system has to do more than merely react to things that have never been in your body before. Things such as food, harmless gut bacteria or dust from the air – all not part of the human body – pose no danger and should not trigger an immune reaction. But as George Bernard Shaw put it in 1930: 'science can never solve one problem without raising ten more problems'.[32] Leaving aside the biggest problem Janeway's ideas faced, which was a lack of experimental evidence to support them, there was a theoretical problem too: germs increase in number rapidly. Just how fast germs multiply is mind-boggling. One virus-infected human cell can produce a hundred new virus particles. This means that just three copies of a virus going through four rounds of replication – taking a few days or so – leads to 300 billion new virus particles.[33] It's not just viruses that behave like this; bacteria divide every twenty minutes in optimal conditions, which means one bacterium can produce 5 billion trillion (5×10^{21}) bacteria in a single day – something like the number of stars in the universe.[34] In practice, germs can't multiply to such an extent within the human body because this level of growth requires an unlimited amount of resources, but even so, germs reach enormous numbers fast; far quicker than our measly average of about two offspring per couple's lifetime.[35] This leads to a crucial problem with Janeway's idea: each time a germ reproduces, it acquires random changes in its genes – mutations – and through these changes it seems likely if not inevitable that some will lose the molecular signature detected by the immune system. In other words, in the whole population of viruses or bacteria, some of them will by chance – because there are so many – have acquired a genetic change which alters the part of the germ which the pattern-recognition receptor was designed to lock onto. Microbes lacking the 'molecular pattern' would escape detection by the immune system and multiply readily.

Janeway realised this and so he predicted that 'the pattern recognised should be the product of a complex and critical [process] in the microorganism'.[36] In other words, the telltale structure of a germ would have to be something so critical to its lifecycle that it would be exceptionally difficult, if not impossible, for the germ to alter it. Janeway had evidence that germs do have features such as these, which are both intrinsic to their survival and also vulnerable to attack, because a feature such as this is what allows penicillin to work. Every time one bacterium divides, it needs to build a cell wall to envelop its two daughter cells. Importantly, the process is so complex that bacteria can't easily change it. Penicillin works by interfering with the last stage of the process. As a result, there isn't any simple genetic mutation that would allow bacteria to escape penicillin's effects. True, bacteria *can* become resistant to the antibiotic by making their cell walls with a very different process, but it's not easy, which is why penicillin remains effective against a huge number of microbes: it locks onto bacterial protein molecules involved in an essential and complicated process.

One scientist recalls that when Janeway presented his paper the audience was 'intrigued but not convinced'. Another recalls that the 'community was not ready for Charlie's thinking'.[37] Standing before many of the world's greatest immunologists, Janeway had the confidence to claim that everyone had missed a hugely important part of how the immune system works, even though, as he put it himself, 'experimental verification … is not available'.[38] Quite simply, at the time, nobody could say if Janeway's ideas were revolutionary or fanciful bunkum.

Janeway's paper was all but forgotten; hardly referred to in another scientific paper for the next seven years.[39] But it touched one person – 4,500 miles away – who would, against the odds, bring Janeway's ideas out of obscurity. In autumn 1992, a student at Moscow University, Ruslan Medzhitov, read Janeway's paper and it changed his life.

*

Born in Tashkent, Uzbekistan, Medzhitov was working for his PhD in Moscow, studying how molecules have evolved to stick to each other, when he read Janeway's paper. The Soviet Union was breaking up and scientific research in the country was in trouble; Medzhitov remembers it as 'a time of huge chaos, with no funds available'.[40] As a result, he could not gain practical experience of working in a lab and had to spend his time thinking and reading, with easy access only to old textbooks that he found quite confusing.[41] Students weren't actually allowed in the library that held a copy of Janeway's paper, but Medzhitov charmed his way in. Browsing the shelves, he stumbled upon Janeway's paper and was immediately gripped by its logic. 'It was one of those moments when a light bulb goes on … it felt visceral … it seemed to explain everything,' Medzhitov recalls.[42] He spent half his monthly stipend on photocopying the paper.[43]

Excited to discuss its ideas further, Medzhitov started sending emails to Janeway. To do so, he was given permission to use the department's email account which had a limit of 300 words per day on account of the costs. Medzhitov recalls how he had to save his message for Janeway onto a floppy disk, which he then gave to the person in charge of the one computer in Moscow University that was connected to the Internet. Any reply that came back would also be copied onto a floppy disk and returned to him that way.[44]

Janeway was proud of his ideas about innate immunity and felt dismayed that they were largely ignored by the immunology establishment, so he was thrilled to receive emails from a student in Moscow wanting to discuss things further. Eventually, Medzhitov asked if he could work in Janeway's lab at Yale. Janeway bounced the idea off his wife but she was sceptical. Medzhitov, meanwhile, won a research fellowship to work for three months at the University of California, San Diego. He borrowed the cost of the flight from a cousin and began working there in 1993, writing software that could scan and organise genetic code – a new area of research at the time. Crucially, he gave a seminar on his work, in broken English, attended by then president of the American Society of Immunology, Richard Dutton.

Dutton was impressed. Medzhitov told him that his fellowship was ending soon and that he was in correspondence with

Janeway by email and he would love to work there. So Dutton left a message on Janeway's answering machine to tell him he thought Medzhitov was a good scientist. And the next morning Medzhitov had an email from Janeway offering him a job.[45]

On 2 January 1994, Medzhitov finally met Janeway face to face. Both were big-picture thinkers, passionate about ideas, and a lifelong partnership and friendship began. The duo's immediate mission was to find out if human immune cells really did have 'pattern-recognition receptors' able to detect telltale signs of germs. One example was all they needed but the problem was formidable, and it didn't help that Medzhitov had so little practical experience. As Roald Dahl wrote in his final children's book, it always pays to 'watch with glittering eyes the whole world around you because the greatest secrets are always hidden in the most unlikely places'.[46] And so it was for Medzhitov, whose eventual success had its roots in an unlikely source: insects.

Like us, insects are also under threat from germs, such as bacteria and fungi, and yet, as scientist Pierre Joly noted in the mid-1960s, insects never seem to suffer from an opportunistic infection. Working in Strasbourg, Joly observed this to be the case even when he transplanted organs from one to another and surmised that insects must have some kind of especially potent immune defence. Joly was joined in his lab by a twenty-three-year-old PhD student named Jules Hoffmann, keen to study insects because his father was an entomologist. Hoffmann set out to understand the insect immunity that Joly had observed and began working with grasshoppers.

When Joly retired in 1978, Hoffmann, then aged thirty-six, became the head of the lab. Over time, Hoffmann shifted the team's focus away from grasshoppers and onto a small fly, drosophila, which feeds and breeds on fruit. Fruit flies were initially used for research in the early 1900s because they're easy to keep, with a simple diet of food scraps, and have a short, two-week lifecycle. Subsequently they have played an enormous role in biomedical research and are at the heart of no fewer than six

Nobel Prize-winning discoveries.[47] But for Hoffmann, another practical reason for switching to fruit flies was that half of his team had become allergic to grasshoppers. Hoffmann's wife Danièle, who had also been his PhD student, was affected especially severely.[48]

Hoffmann's team injected bacteria into the fruit fly and then periodically tested the fly's blood for its ability to kill other bacteria. Once the fly's blood had gained antibacterial properties, Hoffmann knew an immune response had been switched on. His team then set about answering two crucial questions. What kinds of molecules had endowed the fly's blood with an ability to kill germs? And second, which genes controlled the fly's immune response? The first question turned out to be quite easy to answer. Specific kinds of molecules (short pieces of protein, known as peptides) had been identified in silk moths as being antibacterial, and Hoffmann's team found similar molecules in their flies, with different ones being able to kill different kinds of germs.[49] From 100,000 flies, for example, Hoffmann's team isolated the peptide which flies use to kill fungi (nowadays it could be done using about twenty flies).[50]

To answer the second question – which genes are important for a fly's immune response – it turned out that Hoffmann's choice of the fruit fly as the subject of his enquiries was crucial because the insect's genetic make-up was being investigated in other labs for all kinds of other reasons. This separate work gave Hoffmann's team vital clues. One clue was that an insect gene named toll – from the German word *toll*, meaning 'great' – which was important in the development of the fruit fly embryo turned out to be similar to a human gene (called the IL-1 receptor) already known to play a role in immunity. Also, certain genes present in both flies and humans (known as NF-kappa-B transcription factors) had recently been discovered to be important for human immune responses.[51] Spurred on by these recent discoveries, Hoffmann's team set about testing if flies with specific genes inactivated had any difficulty in dealing with infections.[52] Crucial experiments were performed by Bruno Lemaitre, who had joined Hoffmann's team in November 1992.

Over a series of experiments spanning 1993–5, he discovered that flies were dependent on the toll gene to be able to clear fungal infection.[53] This was a spectacular discovery – firmly establishing that genes involved in the embryonic development of the fly were also part of its immune system – and was immediately recognised as such.[54] In September 1996, the front cover of one of the world's most prestigious science journals, *Cell*, showed a striking photo of a fly with an inactive toll gene, covered in a fuzzy-looking fungus.

In June 1992, before this discovery had been made, Hoffmann had travelled to Yale to meet Janeway because, Hoffmann recalls, he 'didn't want to live all his life in an insect ghetto'.[55] These discussions had led to a joint research programme to compare immunity in insects, mice and humans, and, in 1993, Hoffmann organised what was probably the world's first meeting devoted to innate immunity, which was held in Versailles.[56] In spring 1996, at a follow-up meeting in Gloucester, Massachusetts, Hoffmann first told Janeway and Medzhitov about his team's discovery: that the toll gene was important in the insect's defence against a fungus. Janeway and Medzhitov were thrilled.

The precise sequence of subsequent events varies according to who's telling the story. Medzhitov says that he had already been working on a human gene similar to toll for some time, while others suggest that the discoveries in insects *led* him and Janeway to then search for something similar in humans.[57] Either way, Medzhitov, working in Janeway's lab, ramped up work on the human equivalent of the insect toll gene and the important thing is that he discovered that it could switch on the activity of other genes (specifically NF-kappa-B transcription factors) known to be involved in immune responses.[58] Put together, the implication of these discoveries was profound: they showed that life forms as different as insects and humans share a genetic heritage for fighting disease.

Other research teams then uncovered many more genes, in mice as well as humans, like the insect toll.[59] Collectively, they are called toll-like receptor (TLR) genes – named for being a set of genes where each encodes for a receptor protein similar to insect toll – and there are ten in humans. As work progressed each gene was given a number. Medzhitov's original human toll is now called TLR4. Experiments with mutant mice showed that these different toll genes were essential for immune reactions to all kinds of bacteria and viruses. Even so, while it was clear toll genes were *somehow* important for immunity, nobody really knew how they worked. Not until 5 September 1998.

Bruce Beutler, Chicago-born and working at the University of Texas Southwestern Medical Center in Dallas, had spent the last five years obsessed with one thing: finding out which gene was crucial for an immune reaction to occur in mice that had been exposed to lipopolysaccharide, or LPS – the chemical normally found in the outer coating of bacteria, which had been shown to be an especially potent adjuvant. The problem was widely known as important because the gene involved would likely give a big clue as to how this bacterial molecule was sensed by the immune system, so Beutler was in a race with other labs to identify it. He lived and breathed and dreamt about the problem.[60] He likens it to looking for a lost coin in a living room; especially frustrating, because you just never know when it's actually going to show up.

That year, 1998, didn't begin well for Beutler. In April, he was told that his research funding would have to end soon because he had already spent long enough trying to solve this problem. And at home, Beutler separated from his wife Barbara, beginning a long divorce process which would involve a jury trial eventually leading to joint custody of their three sons. 'Tough times at home coincided with the toughest phase of the [genetic] work,' Beutler recalls.[61] As well as leading his research team, Beutler himself analysed the data his lab had obtained, writing his own computer code to help with this process.[62] On the evening of 5 September, he was filled with joy when an analysis on his computer screen indicated that the crucial gene for detecting the bacterial molecule LPS in mice was very similar to Hoffmann's insect toll gene and Medzhitov's human gene, TLR4.

Finally the pieces came together to reveal the big picture: the TLR4 gene encodes for a protein molecule that is able to interlock with a component from the outer wall of bacteria (LPS). In other words, the TLR4 gene encodes for a pattern-recognition receptor, the very type of molecule that Janeway had predicted existed – one of the eyes of the immune system, as Beutler puts it – giving immune cells with this receptor protein protruding from their surface an innate ability to lock onto bacteria. When TLR4 locks onto the bacterial molecule LPS, this signifies that there is something in the body that may very well require an immune response. Beutler says that he wasn't actually inspired by Janeway's earlier ideas directly; he came at the problem from a different view, thinking that the gene which allows the immune system to respond to bacteria is of obvious importance anyway and likely encodes a receptor protein at the surface of immune cells.[63] Besides, Beutler thinks that the days of great thinkers driving biology forward have long since passed – observations drive progress now.[64]

The first person Beutler phoned with news of the discovery was his father – his role model – who was a pre-eminent scientist himself and had always emphasised working on something important rather than settling for mundane details.[65] Though his father had constantly challenged his son to excel, on this occasion he was 'somewhat nonplussed' at his news.[66] Beutler then phoned his long-time co-workers – a different kind of family – and they were truly excited. Later, Beutler also phoned his funding agency but they replied that their decision made earlier that year, to terminate their support for his research, was irreversible.[67]

Beutler's discovery was published in December 1998.[68] Other teams also reached the finish line – coming to the same conclusion as Beutler with other types of experiment – but Beutler won the race.[69] One team – Danielle Malo's research group in Montreal – reported the same discovery about three months after Beutler.[70] Their paper didn't mention Beutler's earlier report but he made them do so in a subsequent correction to their article, which also clarified that Beutler had first presented the discovery at an even earlier scientific meeting. Researchers in Japan also reported the discovery a couple of months after Malo.[71]

Thirteen years later, on 3 October 2011, Beutler looked at his phone and saw an email with the subject line 'Nobel Prize'. It read: 'Dear Dr Beutler, I have good news for you. The Nobel Assembly has today decided to award you the Nobel Prize in Physiology or Medicine for 2011 … Congratulations!' Slightly disbelieving the news, he opened his laptop to check and saw on Google News that it was really true.[72]

He shared the prize with Hoffmann, as well as Canadian immunologist Ralph Steinman, whose work we'll turn to in the next chapter. Many scientists agreed that these individuals and the scientific discoveries they made were deserving of a Nobel Prize. But a month after announcement of the prize, twenty-four eminent immunologists took the unprecedented move of publishing a letter to the world's top science journal *Nature* to say that the Nobel Committee 'should also have acknowledged the seminal contributions' of Janeway and Medzhitov.[73]

Sadly, Janeway had died on 12 April 2003 from lymphoma, aged sixty, and the rules for the Nobel Prize state that it cannot be awarded posthumously. His obituary in *Nature* said that while 'most scientists only dream of contributing to a paradigm shift – Janeway personally initiated one'.[74] Janeway had published over 300 scientific papers and written a leading textbook on immunology. Bill Paul, a renowned immunologist working at the National Institutes of Health, Bethesda, wrote in 2014 that Janeway would almost certainly have won a Nobel Prize had he not died prematurely.[75] His protégé Medzhitov was, however, eligible for the Nobel Prize – and indeed just before the Nobel announcement, he shared another prestigious prize, the 2011 Shaw Prize, with Hoffmann and Beutler – but another rule for the Nobel Prize in Physiology or Medicine is that it can only be given to a maximum of three scientists and Medzhitov missed out. The Nobel Committee would have undoubtedly discussed Medzhitov's work, but records of these meetings are kept secret for fifty years. We may learn more about their thinking in 2061.

It's not that these scientists would anyway celebrate others winning a prize in their research field; they don't get along. There was a deep rivalry between Beutler's team and Janeway's over the discovery of TLR4's ability to 'see' bacteria. Medzhitov says that he made the same discovery in Janeway's lab around the same time as Beutler, while Beutler says that he got there first and that, at the time, Medzhitov's work was incomplete. To this day, Medzhitov refuses to attend a scientific meeting in which Beutler or Hoffmann have also been invited.[76]

Another complication with the award of the Nobel Prize came in December 2011, when Lemaitre, who carried out the crucial experiments in Hoffmann's lab in 1993–5, set up a special website to claim that he had been slighted by the Nobel announcement because he had really done the experiments for which Hoffmann was being celebrated. Hoffmann, on the other hand, says that key to his lab's success was building a team of people with different expertise and experience, many of whom were involved in the work that led to the seminal discovery of the insect toll gene being important in defence against fungi.[77] In 2012, eight eminent immunologists published a letter supporting Hoffmann, saying that he 'has been impeccable in his assignment of credit and support for his co-workers both while they served in his laboratory and in their future independent careers'.[78] Later, in April 2016, Lemaitre self-published a book, *An Essay on Science and Narcissism*, on the idea that 'narcissism is an advantageous trait for succeeding in science'.[79] These sorts of clashes are not infrequent in science because it's very difficult to disentangle the precise impact of each individual who contributed to a discovery and nobody works in complete isolation.

There can be no question, however, that these discoveries deserve the highest level of celebration. This watershed moment in our understanding of the human body has led to more than 30,000 scientific papers being published on toll-like receptors in the immune system in ever-increasing detail. The immediate next step was to find out what kind of germ each numbered receptor is able to see. While TLR4 locks onto a molecule in the outer wall of bacteria (LPS), TLRs 5 and 10 have been shown to lock onto molecules found in parasites, TLRs 3, 7 and 8 detect some types of virus, and so on. The flood of research that followed also revealed that the toll-like receptors are just *one* kind of pattern-recognition receptor; there are many others, with cumbersome names such as nucleotide oligomerisation receptors, C-type lectin receptors and retinoic acid-inducible gene-1 (RIG-1)-like receptors.

Not only is each pattern-recognition receptor able to detect a different kind of germ, each is also positioned differently in the body – strategically located in a place where that germ might be found. TLR4, for example, is located on the surface of many white blood cells, to look out for bacteria including E. coli and salmonella. Another pattern-recognition receptor, RIG-1, meanwhile, is positioned inside cells to look for the telltale signs of an invading virus, such as influenza. A receptor important in detecting fungi such as *Candida albicans*, responsible for thrush, looks out from the surface of immune cells that are adept at engulfing and destroying fungi. One of the teams working out these details is led by Shizuo Akira at Osaka University, Japan; 'a man of few words but many publications'.[80]

Before these discoveries it was thought that innate immunity was merely a broad-brush defence, in the same way that skin, for example, could be thought of as simple barrier against all kinds of germs entering the body. But when a multitude of different pattern-recognition receptors were discovered – each equipped to detect specific types of germs and switch on a response that is appropriate to the threat – it became clear that the innate immune system is far more complex than had been imagined. The innate immune system doesn't just detect the presence of germs but can recognise the type of germ present and direct the immune response accordingly.

Of the 1.5 million species known on earth, something like 98% of them are invertebrates – animals without a backbone – and they survive disease with only this type of defence. That is, their immune system only uses receptors that lock onto telltale signs of germs. For us, this is just one of the ways in which our body detects disease. The system (or subsystem) uncovered here – innate immunity – forms our first line of defence, an immediate response to the presence of germs.[81] An opportunistic fungal infection or a bacteria entering a cut or wound is often dealt with rapidly by our innate immune system. It is only when our innate immune response can't deal with an infection fully that the adaptive immune response – the action of T cells and B cells – becomes important, a few days after the body has been infected. Infections resolved within two or three days are usually down to germs being detected by pattern-recognition receptors and the appropriate reactions they trigger. In fact, it's been estimated – though it's hard to calculate such a thing – that around 95% of our defences against microbes are attributable to innate immunity.[82] Ever since Jenner first vaccinated an eight-year-old boy against smallpox 220 years ago, we sought to understand immunity and yet, until 1989, mankind had been studying just a part – arguably, as little as 5% – of what makes up our immune defence.

At first, the pioneers of innate immunity weren't thinking about the possible medical applications of their discoveries; they were trying to solve a puzzle about how immunity works. Hoffmann thinks it's important to emphasise that the research was curiosity-driven; he subscribes to Louis Pasteur's view that 'there are no such things as applied sciences, only applications of science'.[83] Indeed, there are many examples of huge medical advances coming from left field. One of the best is X-rays. As cosmologist Martin Rees has said: 'A research proposal to make flesh appear transparent wouldn't have been funded, and even if it had been, the research surely wouldn't have led to the X-ray.'[84] However, it soon became clear that there were important medical outcomes from these discoveries in innate immunity; the most pertinent, back where it all started, was to do with vaccination.

It remains unclear how aluminium salts manage to help vaccines work, even though they have been used successfully in hundreds of millions of people since 1932.[85] But what did become clear was that adjuvants are important because they switch on the innate arm of our immune system. As a result, instead of using aluminium salts, adjuvants could be tailor-made to switch on the innate response using the molecules that had been identified as the specific targets of pattern-recognition receptors. Because of this, pharmaceutical companies changed their view of vaccine research from an area that had relatively low prospects for financial pay-off to one that could be lucrative. This, coupled with the efforts of charities such as the Bill and Melinda Gates Foundation – which sponsored research into new adjuvants for a malaria vaccine, for example – has kept innate immunity a hot topic ever since the pioneering discoveries of the 1990s. One early medical success was a molecule, similar to LPS, which was approved in the USA in 2009 for use in vaccines against the human papilloma virus, which causes cervical cancer.[86] (It's possible – I'm just speculating here – that the Nobel Committee waited for these medical benefits to be clear before awarding a prize in innate immunity, at the cost of waiting too long for Janeway to be included. At any rate, the fact that it took a whole twenty years to move from Janeway's ideas to any medical applications highlights one reason why most curiosity-driven research is funded publicly rather than commercially.)

Other medical applications are likely. Beutler and others think it may be possible, in the near future, to help autoimmune diseases with new drugs that block the action of toll-like receptors.[87] Inhibitors of toll-like receptors may also help prevent problems in transplantation, which also result from an unwanted immune reaction; in this case the patient's immune system reacting against the transplanted organ.[88] While medical interventions that act on our innate immune system are still being pursued, these discoveries open up another space where new medicines could act: in how the different subsystems – innate and adaptive immunity – connect together. This is where we shall turn next.

I once asked Medzhitov if he thought Janeway had a particular trait that was important for his ability to have predicted so much, so many years before everyone else.[89] He replied confidently that many scientists have one big idea, which they stick with throughout their entire career. Janeway, however, like all creative people, had many ideas, and above all he was never afraid of being wrong.

2 The Alarm Cell

Our brain focuses on movements or changes, and the flicker of something can make you jump. We've evolved to react like this because it's better to overreact to a flutter in the wind than to miss a real threat. A momentary scare, no harm's been done. But our immune system has to be more careful. Its power must not be unleashed as a mere precaution. Healthy cells and tissues are easily destroyed by overzealous immune cells – as we see in autoimmune diseases like multiple sclerosis or juvenile diabetes, as well as in conditions such as septic shock.

Like his contemporary Charles Janeway, Canadian immunologist Ralph Steinman was puzzled at how an immune reaction got started. But Steinman had a slightly different way of thinking about the problem. The question which he felt most important to answer was: how does the body decide to make an immune response *with the right level of caution*? This was a crucial question, because, he thought, if we knew how the immune system decided when and how it was appropriate to react, we would know how to regulate immunity and tackle the problems that occur when it goes awry, as in autoimmune diseases. As author Arthur Koestler wrote in *The Act of Creation*, 'the history of discovery is full of arrivals at unexpected destinations, and arrivals at the right destination by the wrong boat'.[1] Setting out to solve this important problem about how the immune system works, Steinman's destination would be a monumental scientific discovery: a new type of cell.

Steinman's parents had wanted him to study religion and take over the family business, a store selling everything from appliances to clothing, but Steinman loved science.[2] At the time, scientists had only recently learnt how to isolate different kinds of cells from blood or tissue. The new frontier was to dissect how an immune reaction worked by mixing together different combinations of cells in a culture dish and testing their behaviour. Determined to be at this frontier and inspired by a series of lectures on the 'new cellular immunology' given as part of his medical training at Massachusetts General Hospital, Boston, in 1970 Steinman joined Zanvil Cohn's lab at the Rockefeller University, New York, which already had a formidable reputation for studying immune cells.[3]

For the first couple of years there, Steinman worked on the main topic of the lab at that time – studying how immune cells can engulf molecules from their surroundings[4] – but in 1972, he turned his attention to another problem, which proved to be especially rewarding: the mystery of the *accessory cell*. At the time, the accessory cell was an idea rather than an actual cell, invented to account for an observation that was otherwise hard to explain: that when isolated immune cells (specifically T cells and B cells) were mixed with something known to be able to trigger an immune reaction, nothing happened.[5] Presumably something else needed to be present for these immune cells to react, but nobody knew what was needed or why. The 'accessory cell' was the name used to refer to whatever that something else might be.

It was known that immune reactions happen readily in the spleen. Using T cells and B cells isolated from spleens that he had dissected from mice, Steinman found like others before him that he could not start an immune reaction in a culture dish without adding 'accessory' cells, and in practice this meant adding whatever from the spleen stuck to glass. So Steinman decided to look closely at what, from the spleen, actually stuck to glass. Among the hotchpotch of cells spread out beneath his microscope lens, some caught his attention because they were stellate and spiky-shaped. With numerous fine projections emanating from the main body like branches from a tree, these cells looked quite different from the fried-egg-shaped cells drawn in a school textbook. In fact, they looked different from anything Steinman had seen before.

Though he didn't know at the time, cells like this had in fact been seen before – over a hundred years earlier, in 1868 – by German biologist Paul Langerhans. Aged twenty-one at the time, Langerhans saw stellate cells in skin. He believed them to be nerve cells on account of their unusual shape and published a paper describing them – 'On the nerves of the human skin' – while still an undergraduate student.[6] When Steinman watched the strange cells move, he saw that they could, in his own words, 'assume a variety of branching forms, and constantly extend and retract many fine cell processes'.[7] He had never seen cells move like that. Either nobody else had seen these cells move before, or if they had, they hadn't taken proper notice. It wasn't exactly a eureka moment because Steinman didn't have any sense of what this movement or the cell's unusual shape meant. It was more of a 'wow, that's strange' moment, but Steinman had a hunch that these cells were very important.

A scientific discovery such as this, made by just watching cells down a microscope, doesn't happen as simply as might be imagined. One of the reasons it is so difficult has been strikingly illustrated by two Harvard psychologists, Christopher Chabris and Daniel Simons, who asked volunteers to watch a video of six basketball players – three in white T-shirts and three in black – walking around and passing basketballs between them. Chabris and Simons asked the viewers to count the number of times a basketball is passed between two players both wearing white T-shirts, which takes a bit of concentration.[8] Halfway through the video, which you can watch for yourself online,[9] a woman in a gorilla costume walks into the scene, stands among the players, beats her chest facing the camera, and walks off. Afterwards, the viewers are asked if they noticed anything unusual. Despite the fact that eye-tracking equipment showed that all the viewers had gazed straight at the gorilla for an equivalent length of time, only half had noticed her. This 'perceptual blindness' was even worse when tested on a group of expert radiologists, who were asked to look through computed tomography (CT) scans of lungs in search of nodules, which would appear as bright white circles. While some of the scans also showed pictures of a gorilla that was forty-eight times larger than the nodules the experts were told – and trained – to look for, 83% of radiologists missed seeing the gorilla despite gazing right at her.[10]

These experiments emphasise an important truth: we see with our brains rather than with our eyes. Our brains filter and interpret everything detected by our body's sensory organs and because of this, we often see only what we are looking for and fail to notice the unexpected – even if it's literally as conspicuous as a gorilla walking between people playing with a basketball. Merely in order to *see* these new cells, Steinman had to overcome this human tendency. It may have helped that Steinman hadn't set out with a clear idea about accessory cells that he wanted to test: his approach was just to explore – and the invisible-gorilla experiment suggests it's easier to spot something new if you you're not looking for anything in particular. In a darkened room, looking down the eyepiece of a microscope, there's not much between you and the fragment of nature you're examining. In that lonely space – senses focused – maybe we become more open to the new.

But perceptual blindness was not the only, nor even the most powerful, barrier that could have blocked his way, had Steinman been less of a scientist. One is that any number of interpretations might have led him simply to brush aside what he had seen. Famously, when Galileo looked up to the moon with the newly invented telescope in November 1609 and saw light and dark patches on the moon's surface, he realised that the moon isn't smooth, as had been assumed, but has high mountains and deep valleys – whereas an English astronomer, William Lower, who had used a telescope to look at the moon only a few weeks earlier merely remarked that the surface of the moon looked like the treacle tart his cook had made recently.[11] Steinman could have assumed that the strange-looking cells he encountered were variations of cells already known, or cells that had been affected in some peculiar way, perhaps by the process involved in isolating them. The unusual way the cells moved, for example, might have just been something to do with the glass they were stuck to. (It took around three decades before the technology was available to watch the movement of these cells inside a living animal.)[12] As the scientist who discovered vitamin C, Albert Szent-Györgyi, famously put it, the trick is 'to see what everybody else has seen, but to think what nobody else has thought'.[13]

Steinman's working environment also helped. The head of his lab, Zanvil Cohn, was always hugely supportive. Rockefeller University Press ran its own scientific journal, the *Journal of Experimental Medicine*, and it may well have helped that Steinman could publish his early discoveries in such a prestigious in-house journal. But most important of all was who happened to be working in the lab upstairs. On the fifth floor of the building was, as Steinman himself wrote, 'probably the greatest concentration of cell biologists that have ever worked together in a contiguous space', and among them was a certain George Palade.[14]

Said by another Nobel laureate, Günter Blobel, to be the most influential cell biologist ever,[15] Palade was the man who developed the way in which scientists could look at cells with an electron microscope, which uses electrons instead of light to magnify objects thousands of times better than a regular microscope. In fact, the first pictures of cells taken with an electron microscope had been published back in 1945 by a team – Keith Porter, Albert Claude and Ernest Fullam – also at the Rockefeller University.[16] Palade had joined this team and used electron microscopes to study mitochondria, the compartments inside cells where chemical reactions produce the energy a cell needs. Palade then uncovered, for example, where cells make protein molecules, which is vitally important for our understanding of the process which underlies much of the biotechnology industry, the production of insulin, and so on. These discoveries, opened up by microscopy, were revolutionary because, as historian-scientist Carol Moberg notes: 'At the turn of the twentieth century ... anatomists, histologists, pathologists, and biochemists often disputed the realities of the existence of components in a cell. Many viewed the cell as a bag of enzymes filled with formless protoplasm and devoid of structure.'[17] The Rockefeller University, still a relatively small institution, was the internationally renowned hub where our modern understanding of what goes on inside cells began.

Steinman used Palade's electron microscopes to peer inside his spiky cells. Crucially, this reassured him that they really were different from other kinds of immune cell. They had, for example, much more cytoplasm, the thick liquid which fills the cell beyond its nucleus, than other cells. Confident that they were new, he struggled with what to name them. Deciding on a new scientific name is a rare privilege. Steinman considered naming the cells claudiacytes after his wife Claudia, without whose love and support, he often said, he could not have been as successful in his research.[18] (While having her own career in real estate, she did most of the work raising their son and twin daughters while Steinman was often away from home.)[19] In the end, Steinman settled on the name *dendritic cell* – from the Greek word *dendron* meaning tree, on account of the cell's most obviously distinctive feature, the many branch-like protrusions emanating from its main body.

Though dendritic cells are found throughout the body – in blood, skin and nearly all of our internal organs – they are fairly uncommon in each place. So the next step in what turned out to be Steinman's forty-year quest to work out what these cells do in the body was to find a way to isolate them so that they could be studied in detail. This was no trivial task – it took him five years to hit upon a procedure that worked – and again, those who happened to work upstairs played a crucial role.

Up on the seventh floor, a team led by Christian de Duve were breaking up cells with detergents and other chemicals so that their innards could be separated and analysed. They did this using a centrifuge – an instrument that spins objects (in this case, test tubes full of broken-up cells) around like a washing machine but much faster, hundreds of times a second.[20] The reason this works is that different components of cells have different densities, and denser parts of cells are pulled by centrifugal forces towards the bottom of the test tube, while lighter parts settle (or 'sediment') nearer the top. It's then easy to siphon off fragments of cells, allowing them to be studied separately. Using this method, de Duve's team had been able to identify a wondrous world of *organelles* – literally, little organs – inside cells. The nucleus is the cell's largest organelle, relatively easy to detect, but de Duve found that many other minute compartments fill the cell's interior – small bags, enclosed by membranes, which isolate different reactions and processes. 'I have roamed through living cells, but with the help of a centrifuge rather than of a microscope,' de Duve said when accepting the Nobel Prize with Palade in 1974.[21]

Steinman borrowed de Duve's methods and adapted the centrifuge to separate different types of cell instead of fragments of cells. Those cells that naturally have a different density were easily separated by a few minutes' whirl in a centrifuge – red blood cells are very different from immune cells, for example, and are easily removed in this way. But to isolate dendritic cells, Steinman had to work out a process to make them sediment separately from other immune cells, even those that have a similar density. It took years to work out how to do this by what was essentially trial and error. In the end, the process that worked involved several steps. In a first round of purification, immune cells (including dendritic cells) would rise to the top of the tube spun in the centrifuge, while smaller and denser cells sank to the bottom. The immune cells would be siphoned off and left on glass for an hour. On account of the fact that cells vary in their 'stickiness', according to the types of protein molecules coating their surface, some cells, including dendritic cells, would become stuck to the glass during this time, allowing other types of cells to be washed away. Overnight, the remaining cells would themselves come unstuck from the glass allowing Steinman to subject them to a reaction that clumps immune cells other than dendritic cells around red blood cells. Then followed a second round of centrifugation in which the red blood cells were spun away, taking the rest of the immune cells with them, leaving the dendritic cells behind.

The difficulty of the procedure and the fact that it required specific know-how – in the same way that you can't easily learn to ride a bike just by reading about it – probably helped Steinman in the long run: it meant he had dendritic cells all to himself, without much competition, for at least ten years.[22] But another reason scientists didn't rush to study dendritic cells was that many didn't believe they were a new kind of cell. Most scientists thought that Steinman had isolated a type of cell that had already been discovered – in 1882 – by Ukrainian zoologist Ilya (or Elie) Metchnikoff; a discovery which won him a Nobel Prize in 1908.[23]

Temperamental but widely regarded as a creative genius, Metchnikoff reasoned that 'disease is not the prerogative of man' – animals also get sick – and so it would be informative to watch what happens in animals when they encounter danger.[24] He studied, among many species, starfish larvae which, crucially, are transparent enough to be observed alive under a microscope. In a private laboratory in Sicily he watched what happened to starfish larvae when he punctured them with a sharp splinter. (Legend has it that he pierced them with a rose thorn.) What Metchnikoff saw began a whole new way of thinking about immunity: he saw that some of the starfish's cells *moved* towards the injury.

Perhaps in part because he had recently learnt from a course in pathology that germs were sometimes found inside white blood cells, Metchnikoff had the idea that the cells were moving to the site of injury specifically to engulf – or eat – disease-causing microbes that might enter the wound.[25] 'It struck me,' Metchnikoff said, according to a biography that his wife published after his death, 'that similar cells might serve in the defence of the organism against intruders … I felt so excited that I began striding up and down the room and even went to the seashore in order to collect my thoughts.'[26] Instead of thinking about illness solely from the viewpoint of the afflicted organism, Metchnikoff realised that disease, or at least some kinds of disease, involved two species locked in battle – a battle, in his words, between 'the microbe from outside, and the mobile cells of the organism itself'.[27] He discovered, in other words, that some cells have the specific job of protecting an organism against disease: immune cells. On 23 August 1883, he publicly proclaimed that 'animals disarm bacteria by eating and digesting them'.[28] Later, with the help of colleagues, Metchnikoff named the cells he had discovered as *phagocytes*, and their job of digesting harmful bodies *phagocytosis*, derived from the Greek for 'cell-eating process'.[29] The type of cell best able to eat germs came to be called the *macrophage*, the 'big eater'.

Scientists had, in fact, reported the same process some years earlier, and their work has been largely ignored in the history of immunology.[30] Metchnikoff, however, studied the idea that immune cells can engulf germs in exquisite detail; he compared cells from different species, from different organs, at different temperatures, using different stains, and watched what happened with different types of bacteria. He even studied the effects of narcotics. And humbly, he acknowledged himself that he was not the first to discover the process. His pre-eminence is deserved largely not because he discovered immune cells in one brief moment, watching starfish larvae stabbed with a rose thorn, but because having observed the reaction and formulated an idea of what might be going on, he then persevered in his attempt to understand it.

Likewise, Steinman didn't identify dendritic cells the moment he first saw them down a microscope. This moment was only the beginning – and it would be putting it mildly to say that scientists were sceptical of Steinman's claims at first.[31] One of Steinman's students recalls the reaction to him talking about dendritic cells at an international meeting as simply 'abusive'.[32] Most scientists thought that the cells he had isolated were macrophages because these cells were also known to stick to glass and were more numerous than dendritic cells. For Steinman to persuade the scientific community, not only would evidence be vital but travel would be too. Flights had become cheaper around this time and as a result scientists couldn't rely on publications alone to get their work noticed; it became increasingly important to travel to meetings in person if you were to have your ideas heard and discussed. As a result, Steinman's family vacations were often organised around immunology conferences.[33]

Experiments which Steinman's team carried out in the early 1980s were crucial in persuading the community that dendritic cells were different. A student in Steinman's lab, Michel Nussenzweig, compared the reactivity of T cells when they were in the presence of other immune cells and found a unique potency of dendritic cells to switch on the reaction. In other words, Nussenzweig's work provided strong evidence that dendritic cells were the mysterious accessory cells.[34] As tools and expertise developed so that different types of immune cell could be studied more easily, reagents to mark dendritic cells out from other cells were made[35] and used by Steinman's lab to show that in fact, dendritic cells could stimulate an immune reaction at least a hundred times better than macrophages or any other type of cell.[36] In 1982, another student in Steinman's lab, Wesley van Voorhis, discovered human dendritic cells – all the early work was done with mouse cells – and showed that these too were potent at triggering immune reactions.[37]

Even when Steinman's lab had convinced most scientists that he had discovered a new type of cell, the years of endeavour hadn't got him far towards a clear answer to his original question: how does the body decide to make an immune response *with the right level of caution*? Steinman had discovered that dendritic cells were potent in starting an immune reaction, but he didn't know why, how or what this meant for the working of the immune system as a whole. The path to really understanding the function of dendritic cells opened up only when Steinman and his team found out that the ability of dendritic cells to switch on an immune reaction could change.

One of the people who had an important role in uncovering this was a dermatologist named Gerold Schuler, who joined the team in 1984.[38] Others in Steinman's team had found that dendritic cells isolated from skin were much less potent at stimulating an immune response than dendritic cells isolated from the spleen, but nobody understood why this was, nor what it meant for how dendritic cells work in the body overall. Crucially, Schuler found that when dendritic cells were freshly isolated from the skin, they were indeed quite weak at triggering an immune reaction but when these same cells were cultured in the lab for two or three days, they became potent.[39] This meant that dendritic cells do not exist in just one state; they exist in two states, 'on' and 'off'. The process by which they switch into the 'on' state Steinman called *maturation*, leading to the two states of dendritic cells being called *mature* and *immature*.

As the name suggests, mature dendritic cells are those that are switched 'on' to trigger an immune reaction. Immature dendritic cells are 'off' in the sense of not being able to trigger an immune reaction very well, but what also became clear is that immature dendritic cells are not by any means inactive. They have at their surface many different pattern-recognition receptors, the types of receptors Janeway predicted must exist, such as toll-like receptors, as well as other receptors which endow them with an innate ability to sense the presence of and capture bacteria, virus particles and fragments of infected dead cells from their surroundings. In other words, immature dendritic cells are good at phagocytosis, the eating process. A picture of the two states of dendritic cells therefore emerged in which immature dendritic cells efficiently sense and capture foreign substances in the body, while mature dendritic cells are powerful at switching on other immune cells to react. Still, knowing that there are two states of dendritic cell didn't in itself make clear what is happening in the body; one other discovery was crucial before everything began to make sense.

By the late 1980s and early 1990s, a sizeable international community of researchers were studying dendritic cells, with Steinman an undisputed leader. A series of symposia on the topic began in 1990, which continues today, bringing everyone together every two years.[40] By the time these meetings began, tools had been developed in several labs which could mark the location of dendritic cells and separately identify whether they were immature or mature. Using these tools, dendritic cells were detected in organs such as the skin, lungs and gut, as well as in the spleen and lymph nodes, the small bean-shaped organs found in your neck, armpits, behind the knees and so on, which are filled with immune cells. (These are what you can feel swollen in your neck when you're ill from an infection; commonly called glands, even though technically they aren't.) The crucial discovery made from this line of research was that dendritic cells in tissues such as skin, lungs or the gut were found to be immature – while those in the spleen or lymph nodes were mature.

From this, a narrative for what dendritic cells do in the body finally took shape. Immature dendritic cells patrol almost all of our organs and tissues but especially places exposed to the outside environment, such as our skin, stomach and lungs. These dendritic cells specialise in detecting germs, using the multitude of pattern-recognition receptors they carry. When an immature dendritic comes across a germ, it engulfs and destroys it. Having done so, it then switches into a different state: it matures. The mature dendritic cell makes a beeline to a nearby lymph node or the spleen, a depot jam-packed with other immune cells. There, in the lymph node, other immune cells are presented with fragments of the germs that the dendritic cells have engulfed. The right type of immune cells to deal with the problem then travel out from the lymph node to the site of trouble. All of this movement happens via the blood and the lymphatic system, a specialised system of thin tubular vessels which carry immune cells to lymph nodes, through fluid called lymph which is similar to blood but lacks red blood cells. Dendritic cells travel to a lymph node via lymphatic vessels, while T cells, for example, move out from a lymph node into the body's tissues via the blood.

The body's response to a cut or wound is, evidently, wondrous and complex. First, immune cells move to the problem area, causing redness and swelling: this is our innate immune response, the first line of defence in which immune cells are alerted to a problem because they have receptors at their surface which detect molecules from viruses, bacteria, fungi, or damaged cells.

But as well as this immediate reaction, a complex choreography of immune cells begins so that another level of response can be established, tailored to the precise make-up of the germs that have entered the body: this is our adaptive immune response. This precise and long-lasting immune response begins when dendritic cells arrive at a lymph node and show to the T cells there samples of molecules from the germs that they have engulfed.

The stellate shape of the dendritic cell, with its multiple protrusions, has an explicit purpose here; it allows dendritic cells to connect simultaneously with multitudes of T cells. Recall that T cells have receptors with randomly shaped ends giving them an ability to lock onto all kinds of other molecules.[41] Most T cells won't have the right-shaped receptor to lock onto anything on the dendritic cell. But a few T cells will have the right receptor to lock onto a molecule from the germ it has engulfed. As these T cells have the right-shaped receptor to recognise the germ, they are the right cells to mount a precisely targeted immune response. When a T cell meets a dendritic cell that has engulfed a germ it is able to recognise, that T cell starts multiplying.

One T cell will divide to increase in number at least a hundred- or thousand-fold in the lymph node. (This expansion of cell numbers is why you can often feel lymph nodes swell in your neck when you have an infection.) Killer T cells – 'killer' being their formal scientific name, not just my attempt to spice up the story – move out of the lymph node to the site of the problem, to kill the diseased cells (such as those infected with a virus). Other T cells, meanwhile, called 'helper' T cells, stimulate other immune cells into action. We now know there to be different kinds of helper T cell. Those formally called type 1 help fight bacteria, for example, while others, type 2, stimulate an attack against parasitic worms.[42] Type 1 helper T cells mobilise macrophages, the big eaters, to deal with bacteria, for example. Type 2 cells, on the other hand, switch on a 'weep and sweep' response, in which (without being too graphic) gut cells *weep* mucus, and muscle contractions in the intestine *sweep* out live parasitic worms.[43]

It's not fully understood how the right T cell response is switched on, type 1 or type 2 (and there are others). This remains at the frontier of knowledge.[44] One important process is that dendritic cells switch on certain types of T cell according to the different kind of maturation signal they have received. Parasitic worms, for example, trigger a different sort of dendritic cell maturation, compared to bacteria. Specifically, this happens because different pattern-recognition receptors within the dendritic cell's varied arsenal lock onto different kinds of germs; one detects bacteria, another detects viruses, another for fungi, another for worms, and so on. These pattern-recognition receptors direct the way in which a dendritic cell matures – changing the repertoire of protein molecules that the mature dendritic cell displays at its surface, for example – and this in turn, triggers a specific T cell response.

In short, dendritic cells detect a problem and switch on the right kind of immune response to deal with the threat. In more formal language, they connect our *innate* immune response, the body's instant reaction to germs, to the *adaptive* immune response, which is longer-lasting and more precise, involving T cells and B cells. Other cells in the body, including macrophages, can also do this, but *only* when the body needs to reignite an immune response against germs that have been encountered before. Dendritic cells are crucial for firing up a precise immune response the first time a particular germ enters the body.[45] They are our alarm cells.

If the story ended here, dendritic cells and Steinman's research would already be enough to warrant widespread attention. But this is still the beginning. The role of dendritic cells in the body turned out to be stranger – far less linear – than these initial experiments led us to believe.

*

'My students assume that when well-respected writers sit down to write their books, they know pretty much what is going to happen because they've outlined most of the plot, and this is why their books turn out so beautifully and why their lives are so easy and joyful, their self-esteem so great, their childlike senses of trust and wonder so intact. Well. I do not know anyone fitting this description at all. Everyone I know flails around, kvetching and growing despondent, on the way to finding a plot and structure that work.'[46]

This description by the author Anne Lamott of how a novelist finds a plot might equally be used to describe how scientists find their stories.

The search for the Higgs boson, sequencing the human genome or sending a spaceship to Mars requires an immense amount of long-term planning and paperwork. But working out what a new type of cell does in the body requires a very different approach. This kind of pioneering research is not an exact science. At least initially, there are no clear theories to be confirmed or denied and there are no international consortia or multidisciplinary teams to coordinate. Progress happens thanks to a few individuals following their nose. It's here that the creative process for an artist and a scientist can be much the same. Scientists and authors alike flail around, kvetching and growing despondent, searching for a plot that works.

Steinman discovered dendritic cells without any grand theory as to how they might trigger an immune response; he had no narrative that might have guided subsequent experiments. He'd been given a ball but it was up to him to find out what the game was. He and his team simply had to find out what happens if dendritic cells are mixed together with these or those other cells, in different combinations: do they multiply, die or secrete this or that protein molecule? Does it matter if they are left for an hour or overnight? Do they change their shape, repel or attract, move faster or slow down, get bigger or smaller, make more or fewer protrusions, switch on or off this or that gene?

At first, all the experiments led Steinman, and others, to the view that dendritic cells were crucial in starting a precise immune response. But then, as different conditions and situations were tested, some experiments showed the complete opposite to be true; that the presence of dendritic cells could *stop* an immune response. Just as Steinman thought he had the game sussed, it turned out that he was only at level one and nobody knew how many more levels there were. However much we know, there's always so much more we don't know.

In one of the experiments that seemed to contradict the earlier research, dendritic cells were exposed to protein molecules alien to the body, but not whole germs. Treated this way, we would not expect dendritic cells to trigger an immune reaction: their pattern-recognition receptors would not detect germs and so the cells should stay immature. Indeed, these dendritic cells did not trigger a reaction in other immune cells, but something else *did* happen. Other immune cells that were exposed to these dendritic cells were rendered unable to participate in an immune reaction later, even when germs really were present. In other words, these dendritic cells triggered a state of apathy, or tolerance, in other immune cells, making them unresponsive.

What kept Steinman going when things didn't quite make sense – what keeps all scientists going – is the faith that nature is coherent, that answers exist. We don't give up, we scrutinise the details: understanding how the same cells initiate a reaction sometimes but stop it at others required us to understand the precise mechanism by which dendritic cells interact with other immune cells. Recall that dendritic cells engulf germs at the site of an infection and then, in the lymph node, show to T cells samples of molecules made by the germs. We now know that the way that they do this involves proteins encoded by a handful of especially important genes: the major histocompatibility complex (MHC) genes or, more simply, our compatibility genes.[47] Proteins encoded by these particular genes protrude from the surface of the dendritic cell. They clasp small samples of other protein molecules from inside the dendritic cell, including molecules from any germs that have been engulfed, and put these up for show at the surface of the dendritic cell. T cells examine these samples of protein put up for display, looking for anything that has not been in the body before.

As well as performing this important function, these proteins are special because the genes that encode them – and therefore the proteins themselves – vary from person to person. By and large, we all have the same set of genes – the 23,000 genes which make up the human genome – but around 1% of the genome varies from person to person, such as the genes which affect our hair, eye or skin colour. Importantly, the genes which vary *the most* from person to person have nothing to do with our appearance but are part of our immune system. Variation in these genes gives the proteins protruding from our dendritic cells, presenting samples of what's currently inside those cells, a slightly different shape. This means that we each present a slightly different sampling of proteins made inside our dendritic cells. This is one reason why we each fare slightly differently when faced with any particular infection.

It is worth mentioning that, to my mind, nobody has a better or worse set of these genes overall. The variant which correlates with a better-than-average immune response to infection with HIV also correlates with an increased susceptibility to other illnesses, such as an autoimmune disease. There's no hierarchy in the system. Genetic diversity across our species is essential to our ability to fight all kinds of potential infections, which is, for me, a powerful and fundamental reason to celebrate human diversity.[48]

The detail here which helped us solve the mystery of the dendritic cell's ability both to trigger a reaction but also to prevent one is as follows: when a T cell locks onto something – something that has never been in your body before, presented within a groove of the compatibility gene protein – that alone is not enough to start an immune response. The T cell needs more evidence that an immune response is appropriate. Essentially, every T cell requires *two* signs that there is a problem. The first sign – Signal One being its formal name – comes from detecting a sample of a protein molecule that has never been in your body before. Signal Two comes from what are called *co-stimulatory proteins*.[49] Co-stimulatory proteins are proteins held inside the dendritic cell shuttled out to the cell's surface when that dendritic cell's pattern-recognition receptors have locked onto a germ (and the dendritic cell changes from an immature to mature state). As a result, they are present at high levels only on the surface of dendritic cells that have come into contact with a germ,[50] effectively providing a molecular mark that signifies that a particular dendritic cell has come into contact with a germ.[51]

In other words, the dendritic cell uses pattern-recognition receptors to detect the presence of a germ, or another sign of trouble such as fragments of an infected dead cell, and then the dendritic cell matures (or switches on) and presents samples of that germ to T cells. T cells which have the appropriately shaped receptor to lock onto something presented by a dendritic cell, i.e. something not from the body, require the presence of a co-stimulatory protein on that same dendritic cell as a signal to know it's from a germ, and that a response is needed. If a T cell locks onto something presented by a dendritic cell but doesn't see co-stimulatory proteins, it knows that it is reacting against something *not from a germ*. It may be a molecule that hasn't appeared in the body before for some other reason; maybe it is food or new proteins made during pregnancy or adolescence. In this situation, the T cell doesn't just abort an immune reaction; it switches into another state and becomes a *tolerant* T cell. This T cell is now unable to cause an immune reaction, even at a later moment. In this way, dendritic cells have the power to *switch off* T cells which could otherwise attack healthy cells or tissues.

Scientists studying the immune system often claim that the fragment they're working on is the most important part of the system. Indeed, the system is so complex and layered that equally strong arguments can be made that T cells are especially important, or B cells, or macrophages, or pattern-recognition receptors, and so on. But dendritic cells really do have a special place in the system. They have an ability to switch the immune system on *and* off – both to control our immunity against germs and to stop our immune system attacking healthy cells and tissues. Uncovering the workings of dendritic cells – an endeavour begun by Steinman, but later involving thousands of other scientists – eventually answered his original question as to how the body launches an immune response cautiously: it requires more than one signal before doing so.

*

All along, Steinman was motivated by the belief that his research could be used to design better treatments for disease.[52] Since dendritic cells are absolutely necessary to get an immune reaction started the first time a germ is detected in the body, they are effectively the body's natural adjuvant. We still don't precisely understand how chemicals such as aluminium salts work as an adjuvant, but it is likely they act on dendritic cells, making them switch from an immature to a mature state as if a real germ were present.[53] Surely, Steinman felt, we should therefore be able to use dendritic cells to create new kinds of vaccines against HIV, tuberculosis or cancer.

Japanese scientist Kayo Inaba performed an experiment in Steinman's lab in 1990 which showed that a dendritic-cell-based vaccine could work. At this time, the field was undoubtedly male-dominated; there were, in Inaba's words, 'no women working in immunology', which scared her.[54] (In truth, there were some women working in immunology at the time, but not many.) The experiment she performed is now widely recognised as groundbreaking.[55] First she isolated dendritic cells from mice. She then exposed these dendritic cells in a culture dish either to extracts from tumour cells or to proteins not naturally found in mice. Having been bathed in these other molecules, the dendritic cells were then injected back into animals. Mice given these dendritic cells were then able to mount an immune response against the same molecules the dendritic cells had been exposed to.[56] In other words, she had found that dendritic cells could be switched on outside the body and then injected back into the body to ready the immune system. This was a new way to trigger an immune reaction and, potentially, a new kind of vaccine. In 1992, Inaba moved back to Japan where she broke more new ground: she became the first female associate professor in Kyoto University's Faculty of Science, and at the time of writing, she is a vice president of the university where she actively champions gender equality.[57]

The aim of a dendritic-cell-based vaccine, then, is to use these cells to switch on the body's defences against, say, a virus like HIV, the tuberculosis bacteria, or cancer cells. Inaba's experiments showed how this can work in mice. And as immunologists often quip, that's good news for mice. Testing the procedure in humans is far more complex. In the case of a cancer patient, for example, dendritic cells would have to be isolated or derived from a blood sample, then bathed in a culture dish in protein molecules that were specific to the person's own cancer cells. An adjuvant (components of bacteria, for example) would need to be added to the culture dish to mature the dendritic cells so that they were in a state to activate other immune cells. The mature dendritic cells – having taken up molecules from the patient's cancer cells – would then have to be injected back into the patient. If all went to plan, the dendritic cells would then travel to a lymph node and show to the T cells there samples of molecules from the patient's own cancer cells. In this way, appropriate T cells – those able to detect the cancer – would be switched on and an immune reaction against the cancer would begin.[58]

An idea for a medical procedure of this complexity usually gets tested step by step over many years, maybe even decades. Studies using cells in a lab dish lead to small studies in animals, then larger studies maybe using other animals, followed by small safety trials in humans – a tinkering of protocols at every step – and then finally, the idea is really put to the test in a clinical trial. In March 2007, Steinman suddenly didn't have time for any of that. An advanced cancer – a mass of cells the size of a kiwi fruit – was found growing on his pancreas. He was told that he had just a few months left to live. When he broke the news to his children, he told them, 'Don't google it.'[59]

It's a situation we all dread, but also one that we all mull over from time to time: what would you do if you were told that you only had a short time to live? Some of us would quit our jobs and go and see whatever it is in the world that we haven't yet seen but always wanted to. But Steinman wasn't the type to change his plans. He continued his scientific mission with one difference: he would now experiment on himself.

In setting out to use dendritic cells to cure his own cancer, Steinman hoped that his life's work could save his life. He didn't embark on the new project alone. Friends and colleagues from around the world came together to think up ways in which Steinman's tumour could be tackled. This was to be a huge single-person trial; a disproportionate effort to save a life, born of love and respect for Steinman and his achievements. Every idea was on the table.

It wasn't that Steinman was prepared to subject himself to renegade back-room experiments, injecting himself with bubbling concoctions. Everything still went through the regulators, which meant huge amounts of paperwork for everyone involved. But in the attempt to save Steinman's life, hazards and risks were all examined afresh. Normally, for example, in all labs working with human blood, researchers are taught to never work with their own blood.[60] For Steinman, specific compassionate-use protocols were submitted ad hoc to the US Food and Drug Administration (the FDA). The regulators were responsive and things which could take months to sort out were sometimes turned around in days.[61]

Steinman's first PhD student, Michel Nussenzweig, was by that time a professor at the Rockefeller University, New York. He took some of Steinman's tumour, removed during surgery, and grew it in mice for further analysis. Meanwhile, Ira Mellman, vice president of oncology research at the company Genentech, who had worked with Steinman as a postdoctoral researcher, had his team culture cells from Steinman's cancer, and then tried several drugs on it which he had access to but which had not yet been tested in clinical trials.[62] In Toronto, another of Steinman's friends analysed the specific genetic mutations in his tumour. In Tübingen, Germany, another extracted protein molecules from the tumour to be used in experimental vaccines.[63] One of the scientists who helped knew Steinman from having spent her high-school summers gaining work experience in his lab.[64] Mellman recalls meeting with Steinman in his office to work out what they should try and not try: 'It was a totally natural scientific discussion, except we were talking about his tumour.'[65]

In all, Steinman tried eight different experimental treatments, including three vaccines based on dendritic cells. For two of these vaccines, Steinman's dendritic cells were isolated and modified – in different ways – to contain molecules from his tumour. For one, Steinman's dendritic cells were infused with DNA from his cancer cells. The other approach was to bathe the dendritic cells in cancer-cell proteins. In each case, the dendritic cells were injected back into Steinman's blood – many times over the course of several months – the hope being that they would now be able to trigger an immune reaction against the cancer.

A third vaccine worked differently. For this one, Steinman's tumour cells were isolated and genetically modified so that they would secrete a protein molecule (with the cumbersome name of granulocyte macrophage-colony stimulating factor) which stimulates dendritic cells and other immune cells. The genetically modified tumour cells were then exposed to a high dose of radioactivity, which would prevent them from multiplying as an active cancer does. The tumour cells were then injected back into Steinman's bloodstream. Again, the thinking behind this was that the irradiated tumour cells would attract the attention of Steinman's dendritic cells, be engulfed by them and shown to T cells in a lymph node, so that the immune system would know what to attack.

Steinman also tried more conventional therapies which were in clinical trial at the time, usually in combination with the dendritic-cell vaccines. There was one combination of therapies that he thought particularly promising but it was never tested owing to lack of permission from the FDA. Despite this setback, Steinman always remained optimistic that he was going to be cured.[66] Right up until his last day outside hospital, he was fully engaged in the research, trying to work out ways in which dendritic cells could be used against cancer. Only half-joking, he wanted to publish a paper in the *New England Journal of Medicine*: 'My tumour and how I solved it'.[67] But on

25 September 2011, having had dinner the night before with his wife, three children and three grandchildren, he was admitted to hospital for the last time.

It's impossible to know if any of the experiments extended his life; a single person's case history has no statistical significance. But Steinman was always adamant that they were working. The initial prognosis was that he should survive weeks to months, and the chance of his surviving a year was put at less than 5%.[68] In the end, he survived four and a half years, until 30 September, aged sixty-eight. In all likelihood, Steinman's cancer was already so advanced that even if the experimental treatments did boost his immune system, the tumour cells had likely found ways to avoid being attacked. 'It was a laboratory experiment that worked for a while, we think, but we can't go back and repeat it, so we'll never know for sure,' Mellman said.[69]

Three days after he died, Steinman's wife Claudia got up before sunrise to get a glass of water and saw Steinman's BlackBerry blinking in a bowl next to his keys. It had been left untouched for days, but Claudia saw the message, time-stamped 5.23 a.m.: 'Dear Dr Steinman, I have good news for you …' She shouted up to her daughter, who was asleep: 'Dad won the Nobel!'[70] Looking back, Claudia recalls how her husband 'wasn't there to share the joy … [and it was] very bittersweet'.[71] Hardly anybody knew that Steinman had died when the prize was announced. One acquaintance – and there must have been many others – was embarrassed by sending a congratulation email.[72] Indeed, had the Nobel Committee known of his death, they would not have been able to award it to him. As it was, they received the news an hour after the prize was announced. The Nobel Committee met to decide what to do. If they decided not to give Steinman the award, his share of the prize money, close to half a million pounds, would have likely been added to that of his co-winners Beutler and Hoffmann. In the end, they decided that in these exceptional circumstances, the prize would stand. In the same year that Janeway was denied the prize owing to his death, Steinman was awarded the prize despite it.

Steinman remains the only person ever to have received a Nobel Prize and not know it. He could have – and most scientists would agree, should have – won the prize sooner. Peering down a rabbit hole, Steinman opened up a wonderland of immunity; a world full of peculiar, strangely shaped characters that interact in a complex system, in which many types of cell share information to coordinate their activity in fighting disease. As Mellman puts it: 'Here's a guy who single-handedly started a whole field and stuck with it after the rest of us would have given up to save our careers.'[73]

By the end of his life, Steinman was widely celebrated by the large community of researchers studying dendritic cells. Like a tree known by its fruit, his name will forever be linked with the dendritic cell. Yet like all scientists, he died with some ambitions unfulfilled. All along, he wanted his research to help medically. He achieved only some level of success in this. One dendritic-cell-based vaccine increases the survival of prostate cancer patients by about four months, and has been approved for use in the USA by the FDA.[74] But dendritic-cell vaccines are not yet commonly used to treat cancer. Clinical trials of other dendritic-cell-based vaccines are ongoing, so this type of treatment could become more common in the future, but there are still difficulties to overcome.

One reason why dendritic-cell vaccines are not more effective is that tumours have evolved ways to thwart the immune system. Some tumours, for example, secrete their own protein molecules which prevent dendritic cells from displaying the co-stimulatory proteins at their surface. Dendritic cells affected in this way would not only fail to help; they would actively switch off the body's immune defence – causing T cells to be tolerant to the cancer – potentially making things worse for patients.

A second problem is that when dendritic cells have been switched on outside the body, arming them to trigger an immune response when reintroduced, they tend to lose their ability to migrate within the body. When injected back into a patient, they only rarely make their way to a lymph node where they need to be to meet T cells and switch on an immune response.[75]

A third problem with dendritic-cell vaccines is that, as recent discoveries have shown, there are in fact, many types of dendritic cells. Dendritic cells in the skin, for example, are different from dendritic cells in the gut, which are also different to those found in blood, and there are many variants of dendritic cell even within each location. To some extent, this makes the immune system akin to an ecosystem; cells in different habitats have many similarities but also vary and may adapt if they relocate. Understanding different types of dendritic cell is a current frontier of research. In fact, it is possible that we haven't yet finished Steinman's original experiment, to find the best cell – the accessory cell – for switching on an immune response. There may be a subtype of dendritic cell that is especially potent at triggering immune responses in the context of a vaccine.[76]

Within his lifetime, Steinman's gift to humankind wasn't new medicines; it was a new consciousness of the human body. For centuries, we have known that blood circulates in the body, distributing oxygen and nutrients. Steinman, and the thousands of scientists around the world who eventually worked on dendritic cells with him, unravelled the details of another great dynamism within the human body: that different types of immune cells shuttle between our organs and tissues, into lymph nodes and out again, to defend us continuously and vitally.

Beyond dendritic-cell vaccines, Steinman's broader vision that new medicines could work by harnessing the power of the immune system remains very much in vogue. But a whole other layer of communication within the immune system had to be revealed before this really took hold.

3 Restraint and Control

In the summer of 1956, two scientists met in Mill Hill at the edge of London at the National Institute of Medical Research, the seven-storey building where the influenza virus had been discovered in 1933[1] and which was later used as the fictitious psychiatric hospital Arkham Asylum for the 2005 movie *Batman Begins*. Jean Lindenmann, aged thirty-one, was Swiss and a relative newcomer to scientific research. British scientist Alick Isaacs was three years older and had already achieved an international reputation for his experiments with viruses, having spent three years in Australia working under Nobel laureate Macfarlane Burnet.[2] Lindenmann had studied physics first, at the University of Zurich, but switched to medicine when use of the atom bomb changed his view about what he should do with his life.[3] As a teenager, he suffered from tuberculosis and lived apart from his parents for many years. Perhaps because of this, Lindenmann was quiet and shy. Isaacs, on the other hand, liked to whistle operas for his colleagues to identify.[4] Scientific discussion often works well when one person is lucid and explorative, with the other more restrained and able to funnel the excitement into specific plans for experiments. In this case, the collaborative efforts of these two people with different backgrounds and different temperaments would lead to one of the greatest scientific breakthroughs of the twentieth century.[5]

Before meeting Lindenmann, Isaacs had, for many years, been trying to solve a long-standing mystery about viruses.[6] At least as far back as the nineteenth century, it had been noticed that it was relatively uncommon for someone to be infected with two different viruses at the same time. Charles Darwin's grandfather, Erasmus Darwin, commented on the fact that he had never seen a patient with measles who had smallpox.[7] The mystery – why the presence of one virus seemed to block the growth of another – wasn't studied systematically until 1937, when it was established that monkeys infected with one type of virus, Rift Valley fever virus, were protected against infection with another virus, yellow fever virus.[8] Even for cells growing in a culture dish, when two different viruses were added, often only one grew well.

Mysterious as this was, it wasn't widely thought of as a pressing issue when Isaacs met Lindenmann. The hot topic at the time, especially in the Mill Hill institute, was how flu spreads in an epidemic. Isaacs' research team was focused on this – work on his pet project having shrunk to make way for it – and they had found out, for example, that flu which was rampant in the UK in 1951 involved two different versions of the virus. This, and the work of many others in Mill Hill at that time, was pioneering because it informs the way we nowadays use computers to predict the evolution and global spread of flu, which is essential to how the World Health Organization selects strains for each year's flu vaccine.

Understanding how epidemics spread was, and still is, self-evidently important. It would have been much less clear whether studying why one virus blocks the growth of another was a worthwhile endeavour. Deciding what is important enough to get to the bottom of – which might mean spending years working on it – is the biggest decision that any scientist has to make. Some have a gut feeling about what to pursue, but many analyse the possible causes of an observation and ask if any would be groundbreaking if they proved true. If your computer crashed, would working out why lead to a major discovery? Probably not, so it's best to just turn the computer off and on, and not waste time worrying about what happened precisely.

When Isaacs and Lindenmann met, they soon found themselves discussing why one type of virus blocked the growth of another, because Lindenmann had stumbled across the phenomenon in unpublished experiments he had carried out in Zurich. By the time Lindenmann arrived in London, with a year's salary funded by a Swiss fellowship, Isaacs had already worked out how much of one virus was required to stop another and had demonstrated that one virus could stop the growth of all kinds of other viruses, but the heart of the mystery – *how* one virus stops another – remained. Together they talked about possible reasons. One possibility was that a protein molecule that viruses were known to depend on in order to gain entry to cells got used up, or removed, when one virus entered a cell, preventing a second virus gaining access to the same cells. Another possibility was that a molecule required by a virus in order to replicate might get used up, meaning that a second virus could enter the same cells but would be unable to multiply. They realised that either of these answers would be big discoveries if proved true, because as well as revealing how viruses work, they would expose a way in which viruses are vulnerable. It seemed to them both that this problem deserved more attention than it was getting. And so after discussing the issues over tea, they began experiments together on 4 September 1956. Medical science and their personal lives were forever changed by what they uncovered.

Their now-celebrated experiments involved infecting pieces of membrane from the shell of fertilised chicken eggs with flu virus. But instead of infecting the membrane cells directly with virus, they used virus that had first been mixed with red blood cells.[9] Lindenmann and Isaacs knew that the virus would stick to the red blood cells, which are about 10,000 times bigger, but reasoned that this would not prevent the virus from infecting the chicken membrane cells with its genetic material (sending its genetic material into cells is how the flu virus works to replicate itself). However, once the genetic material had emerged from the virus, its outside coating would remain stuck to the red blood cells. These could then be washed off the membrane, bringing the outside coats of the viruses with them. The red blood cells with virus coats stuck to them could then be tested to see if they could still stop a viral infection when added to fresh chicken cell membrane. If so, they reasoned, it would demonstrate that the outside coat of a virus is what blocks a second infection, as opposed to the genetic material of the virus. The experiment took hours – membrane and red blood cells were left to swish about in test tubes rotating on rollers – and while they waited, Isaacs liked to talk about ideas for more experiments, or politics.

They found that red blood cells that had been coated with virus and washed off from chicken membrane cells could indeed still stop another virus infection. This seemed to fit with the idea that the outside coat of a virus was the important factor for blocking a second infection. But this interpretation completely relied on their assumption that the outer coat of a virus would be left stuck on red blood cells. To check this, they looked at the cells in their experiments with an electron microscope (the same type of microscope Steinman used to get a detailed look at dendritic cells). The pictures were blurred and they couldn't tell whether there were husks of virus left on the red blood cells or not. Worse than this, the electron microscope pictures showed that some of the virus had simply detached from the red cells, probably while the cells and virus swished about on rollers. This worried them. It seemed possible that fully intact virus might have detached from the red blood cells, which was what was blocking a second infection. If so, their experiment hadn't revealed anything new at all. By tackling this worry with a new experiment, they struck gold; actually, something far more valuable than gold.

In order to check whether freely roaming intact viruses were present, they carefully decanted the liquid from their test tubes, separating it not just from the chicken egg membrane but also from the virus-coated red blood cells as well. They then added this liquid to fresh chicken membrane cells, and found that it – or something in it – could also stop cells from being infected. But when Lindenmann and Isaacs checked the liquid, they found it contained very few, if any, dislodged viruses, which meant that they had no explanation for what was going on.

They decided to repeat the experiments without the complication of adding red blood cells. Now they found that the liquid taken from a test tube which contained virus and membrane cells was also able to stop virus from infecting fresh cells. Something in the liquid – just the liquid – stopped viral infections. This was the observation that got them on the right path to making an important discovery, but at the time it didn't feel like anything like a eureka moment because they just didn't know what to make of it. They were flummoxed.

Isaacs suggested that something able to interfere with viruses might have been generated in the liquid, but both scientists were also aware that something less exciting might have happened. If the liquid had turned acidic, for example, maybe that could stop viruses? Or perhaps nutrients had been used up by one virus which stopped a second infection? In amongst the back-and-forth about what they should do next, Lindenmann chose to name whatever it was that was causing the interfering activity *interferon*, to sound like a fundamental particle of the universe, like an electron, neutron or boson. He thought it was about time that biologists had a fundamental particle to work on, as the physicists already had so many. On 6 November 1956, just over two months since they started working together, Isaacs titled a new section of his lab notebook: 'In search of an interferon'.[10] And the hard work began.

It no longer mattered that Isaacs was more of an expert than Lindenmann; in uncharted territory, everyone's a rookie. Like detectives arriving at the scene of a crime, not quite sure what they were looking for, they probed the liquid's capabilities for any sort of clue. They found that heat destroyed its antiviral effect, while storage in a fridge did nothing. These results suggested that the pH of the liquid was not important after all – this wouldn't be affected by heat – but that some kind of heat-sensitive factor was the active ingredient. They tested if centrifugation had any effect. It didn't, which argued against the possibility that a large particle in the liquid was causing viral interference (as anything large would have been spun down to the bottom of the tube). They tested if the liquid could stop different types of virus – and it could. Over time, they ruled out uninteresting and circumstantial explanations and began to grow confident that something as yet unidentified, equipped with the power to stop viral infections, was actively at work; in other words, that there really was an interferon.

Reflecting on this time, Lindenmann later wrote: 'If some people feel attracted to research, it is because its real excitements lie precisely in those phases of groping exploration that must seem tedious and dull to an outside observer. A few vainglorious moments of triumph fade fast; the satisfaction of an intellectually honest effort is more enduring. But it is perhaps just as well that this aspect of scientific work hardly ever comes through in popular writing. There ought to be secret joys that science reserves exclusively to those bold enough or naïve enough to answer its call.'[11]

By the end of February 1957, they decided that they had accumulated enough evidence to warrant writing up their claim of a new cell-derived, virus-induced factor which could interfere with virus replication. The head of the Mill Hill institute they worked in, Christopher Andrewes, was renowned for having discovered the flu virus in 1933, and as a fellow of the Royal Society he helped them publish their results across two papers in the society's *Proceedings*.[12] We now know that the ideas in these two papers were correct, but not many people agreed with them at first.

The trouble began when Lindenmann first presented interferon at a scientific meeting in Switzerland in June 1957. After his talk, a Swiss virologist commented that the idea was contrary to all he had read and had to be rubbish.[13] When Isaacs' and Lindenmann's formal papers came out in October, several eminent scientists, especially in the USA, doubted that they had really discovered a new molecule.[14] The sceptics argued that some of the virus must have contaminated their samples and caused the effects which they had attributed to a new molecule. Rumours spread that the work was fanciful and interferon was given alternative names; the misinterpreton or the imaginon.[15] As is often the case with something new, scepticism was not simply malicious. The early experiments were complicated – cells and viruses were incubated together, liquid siphoned off and reused – and it was open to debate as to what exactly in this process produced the interfering factor. Also, the complexity of the experiments meant it was hard for other scientists to reproduce the results.

Every scientist dreads that their experiments will not be trusted by others. Even worse is if their personal integrity and honesty is doubted. Lindenmann's integrity came under fire when he returned to Switzerland, after his year in London. His former boss, Hermann Mooser, decided that he should have been credited in the interferon papers with Isaacs because, Mooser claimed, the work built upon the unpublished experiments that Lindenmann had done in Mooser's lab in 1955, before the work with Isaacs began. Mooser was highly respected (for his work on typhus bacteria) and his accusation was so disastrous for Lindenmann's career that he had to leave Switzerland. He moved from job to job – a couple of years in Bern, three years in Florida – and then finally moved back to Zurich, only after Mooser had retired.[16]

Mooser died believing that he was a co-discoverer of interferon and he had been treated unfairly.[17] In truth, many labs had carried out experiments which had produced interferon – almost any experiment involving live viruses and cells – but they weren't aware of it. At the time Isaacs and Lindenmann reported interferon, there were various hints within the annals of science that factors might cause immune cells to react in some way.[18] All research happens alongside other research and Mooser did not do enough to be widely recognised as a co-discoverer of interferon.

Isaacs confidently rebuked his claims but he was profoundly disillusioned by the criticism of others. He suffered from depression which sometimes required hospitalisation and medication.[19] As one friend observed, he was 'an imaginative scientist who saw the grand picture … [and was] full of ideas, but when depressed he was difficult to get along with'.[20] Isaacs sometimes talked with his closest colleagues about whether or not the discovery he made with Lindenmann could have, after all, just been due to traces of virus in the liquid which they didn't detect. Perhaps the sceptics were right, perhaps interferon didn't exist.

Ideally, this should have been a matter for investigation with new experiments rather than self-doubt, but despite the object-ive tone of scientific papers, the pursuit of new knowledge is an intensely personal endeavour. Isaacs had a nervous break-down in the autumn of 1958. To most scientists in Mill Hill, he was joyous and passionate, full of enthusiasm and energy, but hidden from view, his life was complicated. As a young doctor, he had married psychiatrist Susanna Gordon, in 1949. Theirs was a happy marriage,[21] but because she wasn't Jewish, Isaacs lost the support of his Orthodox family and his father disinher-ited him.[22] The scientific community was like a surrogate family to Isaacs which meant that its support of his work was excep-tionally important to him.

Intensifying the pressure, interferon wasn't only debated at scientific conferences but also in mainstream newspapers and TV. Not that there was great public interest in Lindenmann and Isaacs solving the long-standing problem of viral interference of course, but everyone realised that if interferon could stop viral infections it could be a new wonder drug. The story was covered by the *Daily Express* in 1957, and then much more widely, including on BBC TV, after Isaacs presented interferon at a reception at the Royal Society in May 1958.[23] Interferon even entered the lexicon of popular culture: in a *Flash Gordon* comic strip from 1960, drawn by Dan Barry, spacemen infected with a lethal extraterrestrial virus are saved by a just-in-time injection of interferon. (In fact, there's a subtle error in the comic strip: interferon was shown to work by bringing down the temperature of the ill spacemen but in reality interferon raises a fever when having its effect.)

The government also kept watch over interferon. Parliament and the Medical Research Council, which funded the Mill Hill institute, were still reeling from the fact that penicillin, discov-ered in 1928 by Alexander Fleming in London, had been devel-oped and patented in the US. The government was bitter about losing penicillin royalties, and if someone else had just stumbled on something big, they were not going to let it slip away again. Placing interferon squarely in that category, someone – it's not clear who, maybe Isaacs himself – described interferon as 'anti-viral penicillin'.

There was great pressure on Isaacs, from the government, the scientific community and the public, to prove that interferon was real, that it could work as a drug and to obtain a patent for it. He suffered deeply from the stress and, unknown to his colleagues, attempted suicide at least twice.[24]

Meanwhile, in Isaacs' lab, twenty-eight-year-old chemist Derek Burke was given the task of purifying the interferon molecule so that its chemical nature and its activity could be more clearly established. 'It is essential to know what interferon is chemically in order to understand fully how it is produced and how it acts in the cell,' Burke and Isaacs wrote in *New Scientist* magazine in June 1958.[25] Isaacs thought this would take Burke about six months and then his ideas would be proved right. But purifying interferon turned out to be a Herculean task. The liquid siphoned off from cells and virus contained minuscule amounts of interferon and Burke filled up twelve notebooks ploughing through chemical processes in his attempts to isolate it.[26] In hindsight, it was hopelessly naïve to think this would take six months. It took fifteen years.

On New Year's Day 1964, long before that work had been completed, Isaacs suffered a brain haemorrhage. The haemorrhage was likely related to an abnormal tumorous blood vessel revealed by an angiogram, but out of reach to surgery.[27] Isaacs returned to work three months later but was relieved from his position as head of a division and instead appointed head of a small research team of himself and two others. After returning to work, he suffered, as one colleague put it, from 'a series of episodes of profound mental disturbance'.[28] A second haemorrhage in January 1967 proved fatal. He was forty-five. A year before he died, he was elected a member of the Royal Society, and after his death a symposium was held in his honour in London, which included two Nobel laureates, Ernst Chain (who worked on penicillin) and Francis Crick (co-discoverer of the structure of DNA).[29] 'The field had lost its patron saint,' one colleague lamented.[30] Isaacs' scientific legacy would eventually be celebrated widely, but he died with it in doubt.

In the last years of his life, a series of small clinical trials of interferon were disappointing and pharmaceutical companies lost interest. Soon after he died, however, the promise of interferon was revived by cancer research. Most cancers have nothing to do with a viral infection but there are a small number of viruses that have been associated with cancer.[31] Ion Gresser, a New Yorker working in Paris, tested in mice whether or not interferon could stop cancer caused by a virus, in the same way it stops other viral infections. His experiments showed that it could. But a bigger discovery came out of his nothing-should-happen control experiment – that is, the version of the experiment that all scientists conduct alongside their main experiment, which is identical in all respects except for the exclusion of the one factor being assessed and is performed with the hope and the expectation that nothing will result from it, thus validating the outcome of the main experiment. In this case, Gresser performed the same test on other types of cancer that have nothing to do with a virus, thinking that these more common types of cancer would not be affected. Unexpectedly, he found that animals injected with all different types of cancer cells survived when treated with interferon. In 1969, he reported that, at least in mice, interferon could cure cancer.[32]

A cure for cancer may be a scientific holy grail but this particular cure was met with more scepticism than celebration. The biggest problem was that Gresser hadn't used interferon as such. Nobody had isolated it yet, so he could only use an impure biological liquid siphoned off from cells and viruses, like cream skimmed from milk, allowing others to argue, as they did with Isaacs, about precisely what the active ingredient was. Gresser recalls a colleague trying to comfort him by saying that one day other scientists would repeat his discovery – and forget that he ever did it first.[33]

Separately from these provocative experiments, a passing observation of Gresser's happened also to influence the course of progress towards our understanding of interferon. In one of his lower-profile research papers, published in December 1961, Gresser noted that, like other cells, human white blood cells mixed with viruses also led to the production of interferon.[34] He speculated that this might play some role in the body's immune defence against viruses and suggested that the production of interferon might be used as a diagnostic test for the presence of a viral infection. This caught the imagination of a Finnish scientist, Kari (pronounced 'Kory') Cantell. A loner who preferred to avoid popular research areas, Cantell reasoned that although most human cells mixed with viruses would lead to the production of interferon, perhaps human white blood cells are especially good at making interferon, and if so, these cells could be used to produce interferon in large amounts in the lab. It was a good idea, but it wouldn't have led to anything if luck hadn't played its part next.

Cantell tested his idea on a virus that he happened to have in his freezer called Sendai virus, which is a little like the flu virus, named for the Japanese city where it was discovered. We now know that Sendai virus is especially effective at getting white blood cells to make interferon. Had he used another virus, or even a different strain of the same virus, his first experiment would have failed and he might never have persevered.[35] As it happens, in his first experiment – begun on 8 May 1963 – white blood cells produced ten times more interferon than any other type of human cell he tested. That's not to say it was all down to luck. Cantell insists that having a permanent job at the outset was also important, otherwise his work would not have been funded for the length of time it ended up taking him to purify interferon after this first experiment – nine years.[36]

The complexity of the process he finally worked out gives a sense of why it took him so long. The process relied on the fact that different protein molecules will solidify out from a solution (i.e. precipitate), at different levels of the solution's acidity. Cantell found that he could extract interferon by stirring an initial crude preparation in cold acidic alcohol and then slowly raising the pH of the liquid by adding other chemicals. Impurities came out of the solution quicker than interferon, and could be removed by centrifugation. The whole process had to be repeated several times. Cantell recalls that chemists thought this process very unusual: 'But I was not a chemist, and my ignorance gave me freedom from prejudice.'[37] It had been fifteen years since Isaacs and Lindenmann reported interferon and, just when public interest in the topic was at a low point, Cantell found a way to purify it, opening the way to put their thesis to the test once and for all.

A handful of cancer patients were treated with Cantell's interferon and encouraging anecdotes spread among clinicians and scientists. The public, meanwhile, were thrilled by the idea of a medicine that is produced in and by the human body itself. It fitted with a desire for remedies to be more natural than, say, radiation therapy.[38] In truth, defining any medicine as 'natural' is difficult in a rigorous scientific or philosophical sense because every treatment is an intervention, and all medicines are derived from nature at some level. The US cancer research community were also thrilled by these early results with interferon because they were under pressure to deliver new medicines after President Nixon signed the 1971 'war against cancer' Act.

Swiss-born US virologist Mathilde Krim, whose media tycoon husband was influential in the US Democratic Party, lobbied for funding for interferon research. Krim was well connected with government administrators, pharmaceutical companies and other researchers.[39] Several prominent US patrons for science, including Mary Lasker and Laurence Rockefeller, lent their support to interferon research largely because of Krim's campaigning.

But all clinical-grade interferon was produced in Finland, controlled by Cantell, and for anyone wanting to test interferon in patients there were two hurdles to overcome. One was raising enough money to buy interferon from Cantell. The second was getting him to agree to sell it. He was bombarded with requests from scientists and clinicians as well as wealthy individuals hoping to save their own lives or lives of loved ones.[40] *Time* magazine called Cantell 'stubborn'[41] but he had to be, because supplies were limited, and he wasn't prepared to distribute interferon to whoever happened to be the highest bidder.

Oncologist Jordan Gutterman, working in Houston, Texas, obtained $1 million from Mary Lasker's foundation to test interferon on cancer patients. Gutterman first met Lasker after she asked the head of a cancer centre to line up a few talks for her on the hottest research.[42] Gutterman, at age thirty-six the youngest speaker in the line-up, caught Lasker's attention – he was, and still is, a great explainer – and a friendship began.[43] To persuade Cantell to sell him interferon, Gutterman flew to Arlanda airport, Stockholm, to catch Cantell as he was passing through to give a lecture.[44] The effort paid off and he promised Gutterman the interferon he needed – with a 50% discount.[45]

Lasker was keen for Gutterman to test interferon on breast cancer patients because a close friend of hers had the disease and hadn't responded to other treatments.[46] Gutterman's first patient had said that, because of her tumour, she felt bad that she hadn't been able to move her arm to comb her hair. After treatment with 3 million units of interferon on 12 February 1978, Gutterman looked into her room and saw her combing her long grey hair. In these first tests, five out of nine women, whose other options had run out, showed a partial regression of their tumour.

He later found that interferon helped six out of ten people with myeloma, a cancer arising from immune cells in the bone marrow, and six out of eleven with lymphoma, cancer that affects the lymphatic system. Other scientists reported similar good news, albeit in just handfuls of patients. Side effects were common – fever, chills and fatigue – but these were minor compared to those associated with other cancer drugs. In August 1978, the American Cancer Society gave what was, at the time, its largest ever award – $2 million – to Gutterman for testing interferon. In July 1979, *Life* magazine proclaimed that interferon was all but certain to be a new wonder drug.

In truth, the early tests of interferon weren't rigorously controlled because there wasn't enough interferon available for a proper clinical trial: nearly all of the world's supply of interferon still came from Cantell in Finland. Others found it hard to reproduce his success in isolating interferon because his method was full of small tricks optimised through hundreds of trials and errors. Cantell never patented his methods; he had no wish to make any personal profit, one reason being that he thought his own financial gain would be inappropriate given that his research had been paid for by public funds. But he knew, for example, that using particular round bottles led to greater production of interferon and it was hard to convince others that such nuance was important – at least until they had failed a few times first.[47]

The situation began to change in March 1978 when Cantell took a call from Charles Weissmann, from the University of Zurich, whom he didn't know. The revolution of genetic engineering was in the air, the biotech industry was expanding. San Francisco-based company Genentech had just shown that a human gene could be inserted into bacteria, and these genetically modified bacteria would then produce the human protein encoded by that gene.[48] This works because the chemical machinery which makes proteins inside cells is essentially the same in bacteria as it is in us: bacteria treat an inserted human gene just as they would any other gene and produce the protein that the gene codes for. In 1982, Genentech would make it big when the FDA approved the sale of the first genetically engineered medicine, human insulin.[49] On the phone to Cantell in 1978, Weissmann talked about the genetic engineering revolution, and how he planned to isolate the interferon gene and use it to produce the protein in bulk. It sounded like science fiction to Cantell – and it almost was. Cantell was cautious about working with Weissmann, but Weissmann invited himself over to Helsinki two days later so that he could explain his plan in person. And then Cantell was won over.

Weissmann explained that when Cantell coaxed white blood cells into making lots of interferon, the process must involve an increase in the activity of the interferon gene, and that they could exploit this to isolate the gene. When a cell makes a protein molecule, such as interferon, it isn't made from its gene directly. First, the gene for that protein is copied into RNA, a chemical very similar to the DNA of the gene itself. The RNA version of the gene is often trimmed or modified (the final version being called the messenger RNA or mRNA) and it then leaves the cell's nucleus to act as a template from which the cell makes the protein. If lots of a particular protein is needed, the cell makes lots of copies of the corresponding RNA template. This was the fact that Weissmann's team exploited to isolate the interferon gene. They first isolated the messenger RNA from white blood cells treated by Cantell's methods, knowing that much of this RNA would be for interferon (others would be for other proteins the cell was making). Then, to isolate the RNA specifically for interferon, Weissmann's team injected the different RNA into frogs' eggs and picked out those which produced interferon. The team then used enzymes to convert the RNA template back into DNA, ending up with the interferon gene.[50] The gene was then inserted into bacteria to produce the interferon protein in copious amounts.[51] Each of these steps was tough, at the frontier of biotechnology, and the long hours required meant that Weissmann kept a sleeping bag in his office.[52]

Weissmann led the work as an academic-entrepreneur and co-founder of the biotech company Biogen.[53] At first Cantell didn't realise that his collaboration with Weissmann was part of a commercial enterprise, but looking back, he says that he probably would have gone ahead with it anyway had he known.[54] In fact, all sorts of financial dealings, which Cantell was unaware of, were happening behind the science. The US pharmaceutical company Schering-Plough paid $8 million to part-own Biogen when the company was close to bankruptcy.[55] $8 million appeared to be a bargain when Biogen announced at a press conference on 16 January 1980 that they had produced interferon from genetically modified bacteria and the stock price of Schering-Plough jumped up by 20%.[56]

While the stock market was elated, the science became complicated (as it tends to). Interferon disappointed delegates at the 1982 American Society of Clinical Oncology when only a small percentage of treated patients showed a partial decrease in tumour size.[57] Many drugs look hopeful in a handful of patients only to fail when tested more carefully on larger numbers of people, perhaps partly because the few people tested initially are cared for especially well or are unwittingly selected so that they do better than average. In November 1982, another problem with interferon surfaced: it became all too clear that the side effects of using interferon weren't as benign as once thought. Four patients treated with interferon in Paris died of heart attacks.[58]

By 1984, the consensus was that interferon was not going to be a cure for cancer in any simple way.[59] Some cancers responded better than others, and in 1984 interferon was approved for use in a particular type of leukaemia, but most successes with interferon were partial or didn't last.[60] By this time, it was also clear that there wasn't just one type of interferon. Tadatsugu 'Tada' Taniguchi, working in the Japanese Cancer Research Institute, had isolated an interferon gene from skin cells which was different to the one Weissmann isolated from immune cells.[61] And several different teams discovered that interferon wasn't the only type of protein molecule able to influence immune cells. Beginning in 1976, a series of international workshops set out to classify the various protein molecules of this type that different labs had found. The first meeting was held in Bethesda, USA, and the second, in 1979, in Ermatingen, Switzerland.[62] At first, this was a small backwater of immunology[63] – the mainstream focused on how specific immune responses were triggered – but in time, a new understanding of human biology emerged from these workshops.

There is much that bears witness to every person's existence but one of the especially great reasons to be a scientist is that what you leave behind includes new knowledge. Lindenmann died in 2015, having lived nearly twice as long as Isaacs, but for both lives, long and short, their one year together, the discovery of interferon, is a huge part of what they left behind. Their labour endures because so many others built upon it. Author Margaret Atwood once wrote: 'In the end, we'll all become stories.'[64] Lindenmann and Isaacs are scientific heroes because their story has become an origin story.

Eventually, the existence of interferon opened the world's eyes to a whole host of soluble proteins like it which are in the body for the same purpose: communication between cells and tissues and coordination of the immune system. We now know that there are over a hundred different proteins like interferon, some of which have been studied across thousands of labs while others have been discovered only recently. Collectively they are called cytokines; they are the immune system's hormones. Our immune cells bathe in a cacophony of cytokines – some switch the system on, others turn it off, many nudge its activity up or down a shade.[65] Their purpose is to shape an immune response to fit the type of problem, say a viral or a bacterial infection, and connect the immune system to other body systems. Their actions are incredibly complex – there are cytokines that regulate the cytokines – but as we shall now see it is hard to overstate their importance in how the body works or their potential for new medicines.

*

All human cells can be invaded by microbes and this is often damaging: many viruses, like influenza or polio, kill their host cell once they've multiplied (often just as they leave to infect another cell). Other viruses, like hepatitis B, keep their host cell alive but cause havoc by upsetting the normal chemical reactions of the cell, and a few types of virus can cause cells to become cancerous. To defend against this, almost all human cells can sense when they have been invaded by a germ, using pattern-recognition receptors to detect their telltale signs. As we have seen, some types of pattern-recognition receptors detect a germ by locking onto a molecular shape which is alien to the human body, such as the outer coating of a virus or bacteria. Other pattern-recognition receptors detect the presence of a germ because they lock onto molecules, such as DNA, which are not alien to the body but are in a location where they shouldn't be, giving away that they are part of an invading germ. Dendritic cells have a vast array of different pattern-recognition receptors, which makes them especially adept at detecting different kinds of invading germs, but almost all cells in the body have some types of pattern-recognition receptor. When any cell's pattern-recognition receptor locks onto the telltale sign of a germ, this triggers the cell to start producing interferon. In this way, almost any type of human cell can be induced to produce interferon when, for example, it is infected with a virus.

Interferon turns the infected cell, and other cells nearby, into a defensive mode. It does this by switching on a set of genes appropriately called the interferon-stimulated genes.[66] These genes produce proteins which help stop bacteria and other germs, and are especially potent at dealing with viruses: they can block viruses from being able to enter nearby cells, stop viruses already inside cells from getting into the nucleus of cells (where they need to go to replicate), and prevent viruses from usurping the cell's machinery to make the proteins needed for new copies of the virus. A protein called tetherin, just one of the proteins made from interferon-stimulated genes, thwarts the spread of disease in the body by grabbing onto a virus, such as HIV, just as it tries to leave one cell to infect another.

In the case of some viruses, this response – our innate immune response – is enough to keep the infection under control, but often this only dampens an infection for a few days until our adaptive immune response – led by our T cells and B cells – develops to eliminate the problem completely and provide long-lasting immunity. One reason that an interferon-stimulated response often can't wipe out an infection is that viruses, and other types of germs, counteract its effects. HIV, for example, can destroy the tetherin protein, so that the virus is free to move away from one cell and infect another.[67] Showing how important this is, one out of the ten genes that make up the influenza virus – in other words, 10% of all that it is – is devoted to counteracting the effects of interferon. We once thought that the positions of stars affect our health, but the truth is even more fantastical: our body is locked in an everlasting arms race with minuscule germs.

We each respond to germs in the same way, but only to a first approximation. One reason that some of us are more likely to suffer especially badly from a flu infection is because of a variation in our interferon response genes. About 1 in 400 Europeans has, for example, a non-functional version of one of the interferon-stimulated genes called IFITM3.[68] Normally, the protein made from the IFITM3 gene interferes with how the influenza virus enters cells, though precisely how is not yet understood. (We know that this same gene is used by animals, as mice genetically modified to lack the gene are more susceptible to flu infection.) And people who have a non-functional version of this gene simply lack this component of our immune defence against the virus. In 2012 the non-functional form of this gene was found to be especially common in people hospitalised by an influenza infection. Those in intensive care were seventeen times more likely to have the defective gene.[69] Variations of this gene are also particularly common in Japanese and Chinese people.[70] Because of this, Japanese and Chinese people may be at higher risk for developing severe illness from flu, but this remains to be tested directly.[71]

However, most people with a dysfunctional IFITM3 gene will still be able to fight off a flu infection without a problem, as it is one of many components of our immune response. In fact, it may even be beneficial to lack a functional IFITM3 gene in other illnesses, such as those conditions in which an immune response is the cause of the problem. Indeed, the fact that there is such high prevalence of this genetic variation in Japanese and Chinese people might suggest that there is some situation which is more common in that part of the world in which this variation confers an advantage.

Although we don't yet fully understand this situation, there are at least two different ways in which we can exploit what we do know. Firstly, we could prioritise people for flu vaccination based on their genetic make-up, targeting those who are at higher risk of responding badly if they should get infected. Right now, we don't screen genes routinely, and it may not be cost effective to increase our capacity for genetic analysis solely for flu, because it is cheaper to offer vaccines to everyone anyway. But this may be what's to come, when genetic analysis is more commonplace. Peter Openshaw, who leads influenza research at Imperial College London, thinks it may be especially useful to screen people with Chinese or Japanese ancestry for whom variations in the IFITM3 gene are common.[72]

The second way that this knowledge is useful is by indicating a way to tackle influenza without a vaccine – which may be needed in an unexpected flu pandemic – by boosting our interferon response. This has already been shown to work in mouse cells: the amount of protein made by the interferon response gene IFITM3 has been increased in mouse cells (by suppression of an enzyme which normally limits it) which in turn increased the defence that mouse cells had against an influenza infection.[73] While mice may celebrate, this idea is not yet medically useful for humans because we don't yet know a way to boost the human IFITM3 protein.[74] Further knowledge is needed here.

Although interferon never lived up to its early hype as a cure for cancer, it is important in the treatment of melanoma and some types of leukaemia, usually given as an injection several times a week.[75] As of July 2015, there were still over a hundred open clinical trials testing the use of interferon against a range of cancers.[76] The chief reason that interferon doesn't work as well as we once hoped is that it doesn't stop cancer cells directly. We know now that most, if not all, of the way interferon helps fight cancer is by stimulating our immune system. The problem is that cancer cells are not easy for our immune system to detect – after all, they are the body's own cells gone awry rather than being alien to the body – and so there's a limit on how great an immune response boosted by interferon can be.

There are many different types of interferon – at least seventeen – produced by different cells in the body.[77] Most of our cells can produce the type of interferon that Lindenmann and Isaacs discovered – nowadays referred to as interferon *alpha* – to limit the spread of an infection.[78] Today, interferon alpha forms part of the treatment for hepatitis B and C infections.[79] Other forms of interferon are more specialised: interferon *gamma*, for example, is mainly produced by some types of white blood cell in order to amplify an ongoing immune response. The genes switched on by each type of interferon are being catalogued in an ever-expanding online database.[80] Many of the other cytokines, discovered after interferon, are called interleukins, so named for being the proteins that act *between* (inter-) leukocytes, a formal name for white blood cells.[81] Abbreviated to IL, each type of interleukin is assigned a number, IL-1, IL-2, IL-3 and so on, currently up to IL-37.[82] Like interferon, some of these cytokines also come in slightly different versions (IL-1 has an alpha and beta form, for example), and some differently numbered cytokines have features in common, so that there is an IL-1 family which includes IL-18 and IL-33, for example. The actions of all these cytokines in the body are wondrous.[83]

Each has a multitude of specific effects and here's just one example: IL-1 acts on, among other cells, neutrophils, which are the most abundant immune cells in the bloodstream. Neutrophils are recruited to a cut or wound within minutes.[84] They can engulf germs and destroy them directly. But one of the especially wondrous things that neutrophils do for our defence is that they shoot out a sticky web, or net, made from strands of DNA and proteins, to capture germs moving by.[85] Think Spider-Man, but on a minuscule scale of cells and germs. These webs contain antimicrobials which kill the captured germs. Neutrophils have a short lifespan, just a day or so in the blood, but at the site of an infection, the cytokine IL-1 increases their lifespan dramatically so that they can battle on, shooting out webs and killing germs for up to five days.[86]

To take a second example, IL-2 has a dramatic effect on other white blood cells, such as Natural Killer cells, a type of white blood cell that is especially adept at killing cancerous cells and some types of virus-infected cells.[87] (I've written about these white blood cells in detail in my first book *The Compatibility Gene*.) In my own lab we often use IL-2 to switch on these cells, having isolated them from blood. It's easy to see the effect IL-2 has, just by watching them under a microscope. When IL-2 is added, these cells elongate from a sphere into a Y-shape and change from being inactive in the culture dish to literally crawling about, the front end of the cell pushing against the surface of the dish while the rear part lets go, propelling the cell forward, probing for diseased cells to attack. If a Natural Killer cell meets a diseased cell, a cancer cell or a virus-infected cell, for example, it will latch onto it, flatten up against it and, within a few minutes, will kill it. The white blood cell then detaches itself from the debris of the dead diseased cell – which looks like a bubbling mess down the microscope – and searches for others to attack.

One of the cytokines which turn off immune responses is IL-10. Discovered in 1989, isolated in 1990 and studied by thousands of scientists since, we now know that this cytokine helps protect the body against unwanted immune reactions.[88] IL-10 curbs inflammation when an infection has been eliminated and signals for the body's healing process, the repair of damaged tissues, to begin. IL-10 is also important in our gut, where it keeps immune cells in a relatively inert state to prevent unwanted reactions against harmless bacteria. Mice genetically altered to lack IL-10 suffer from an inflammatory bowel disease.[89] In humans, an overreactive gut immune system can cause Crohn's disease and ulcerative colitis, which affect more than 300,000 people in the UK.[90]

Our knowledge of cytokines leads to a big idea for medicine: to manipulate their levels in the body in order to boost the immune system to fight infections or cancer, or dampen immune reactivity as treatment for an autoimmune disease. As we've seen, boosting the immune system with interferon has been partially successful, but there are multitudes of other cytokines that can be tried. One pioneer – some say *the* pioneer[91] – in boosting the body's immune response to cancer is Steven Rosenberg.

Rosenberg became the chief of surgery at the National Cancer Institute, Bethesda, USA, on 1 July 1974, at age thirty-three, overseeing nearly a hundred staff and an annual budget of millions of dollars. He has stayed there ever since – because he feels it is 'the ideal place in which to do solid basic science and take it to the bedside'[92] – and has co-authored over 800 scientific papers.[93] In a wonderful book, *A Commotion in the Blood*, about the pioneers of immune therapies, science writer Stephen S. Hall calls Rosenberg 'immensely confident'. 'To some,' Hall says, 'he may have pushed too far and too fast; to others, he was just the right person to do the heavy lifting in a field with a reputation for sacrificing far too many mice in order to save far too few human lives.'[94] Rosenberg says of himself: 'I know I am focussed. Perhaps focus is another word for ruthlessness'.[95] He is careful with his time: 'I rarely attend large conferences; if I go at all it is to give a talk and leave.'[96]

His focus – 'I want to cure cancer and everybody who ever has it' – stems from a patient he met while finishing his surgical training in 1968.[97] Twelve years earlier, the patient had a large tumour removed but the surgeon had found other tumours which he couldn't remove. The patient was told that there was nothing more to be done; he was sent home to die. Yet, here he was, a dozen years later, talking to Rosenberg. Rosenberg could easily have dismissed the case as a mistaken diagnosis but he investigated the records, checked the patient carefully, and scrutinised the old microscope slides he found in the hospital's storage.[98] There was no mistake. This person had had several large tumours and, without any treatment for them, miraculously recovered. Spontaneous regression of metastatic cancer is one of the rarest events known to medicine.[99] How did this happen? 'The single most important element of good science,' Rosenberg wrote later, 'is to ask an important question. This was an important question.'[100]

He reasoned that it had to be because of the patient's immune system. In his first attempt at a cancer therapy, he took blood from the patient and gave it to another, an elderly veteran dying of stomach cancer, to see if it helped. The veteran patient joked that he had been taking long shots all his life and that he was due for one to come good.[101] But it didn't work, and he died within two months. Rosenberg himself later acknowledged that the idea was naïve – 'almost embarrassing in its simplicity'[102] – but also that he just had to try.[103]

Next, Rosenberg tried all kinds of different experimental therapies. One involved isolating immune cells from a patient's blood, growing them in the lab to increase their numbers and then infusing them back into the patient's blood. To culture immune cells in the lab, Rosenberg built upon the discovery that the cytokine IL-2 could be used to stimulate human immune cells to multiply. The precise conditions in which immune cells could be cultured had to be worked out by trial and error – even today, culturing human cells is as much craft as it is science. And even with large numbers of immune cells, it wasn't obvious that they would survive when added back to a patient or retain any ability to kill a patient's tumour. There were so many unknowns in what Rosenberg was trying that most scientists wouldn't even have entertained the idea of such an ambitious programme, but no matter how many ifs and buts there were, Rosenberg was driven by one thought; that *if* things did work out, he would be able to cure cancer.

None of Rosenberg's first sixty-six patients was saved by his experimental therapies. Then, in 1984, his sixty-seventh patient, Linda Taylor, walked in. A navy officer aged thirty-three, she came to see Rosenberg with metastatic melanoma, a cancer that attacks the skin as well as other organs.[104] Two years earlier, she had been diagnosed after a mole appeared to bubble over, and she had been given seventeen months to live. She had tried an experimental dose of interferon, but it hadn't helped.[105] Tired of treatments which didn't work, she thought that travelling to Europe might be the best use of her time left. But her family pushed her to keep fighting – and to see what Rosenberg could do.

Rosenberg treated her with infusions of her own white blood cells and several high doses of the cytokine IL-2.[106] The treatment was far from easy; three doses of IL-2 per day and immune cell infusions every two or three days. Taylor vomited often, felt too weak to see her family and she had trouble breathing. One time she stopped breathing. Her pulse collapsed to just twenty beats a minute and she survived only with an emergency resuscitation. Rosenberg pushed against what Taylor's body could take because he was determined to find out if his experimental treatment held any promise. After sixty-six failures, he knew that his efforts would have to end soon if he didn't see any sign of success.[107] He gave Taylor far higher doses of IL-2 than had previously been administered to anyone.

Two months later, Taylor told Rosenberg that she felt her tumours were disappearing. And she was right. Her tumours had died and the clumps of dead tumour cells were being cleared away by her body. Everyone thought that her tumours would probably come back. But they didn't. Taylor had been cured of cancer.[108]

The feud between humankind and cancer is special. One person being saved by a new kind of medicine is international news. Taylor's story fronted newspapers around the globe.[109] Wisely cautious, Rosenberg tried to play things down with the *New York Times*: 'this is a promising first step'.[110] In his auto-biography, he described how he felt at the time as 'satisfied. Not triumphant or vindicated ... Satisfaction speaks to something deep inside oneself, a fulfilment, a peacefulness and fulfilment deeper than triumph can reach'.[111] Thirty years later, Taylor went back to visit Rosenberg and the moment was filmed for a TV documentary. They hugged, emotions welled up and she said, 'I never cry, except around you.'[112]

Trials with larger numbers of patients showed that IL-2 was the important ingredient in Rosenberg's treatment, not the immune cells.[113] But alas, it soon became clear that IL-2 is not a wonder drug. Less than a year after Rosenberg's success with Taylor, another patient given high doses of IL-2 died. This patient had many tumours – twenty in his liver – and only had months to live, but it was Rosenberg's experimental treatment that killed him. IL-2 hindered his body's normal immunity against a bacterial infection and caused fluid to partially fill his lungs. 'It was a dark time,' Rosenberg wrote later.[114] The patient's mother didn't blame Rosenberg but wrote letters to him about her son's life. To Rosenberg, brought up by religious Jewish parents, the gesture reminded him of something he had learnt about the Holocaust: that 'those who suffered in it had most feared not being remembered'.[115]

IL-2 seemed to offer patients either spectacular success or tragedy – and neither Rosenberg nor anybody else could predict which it would be. Various clinical trials, large and small, have since proven that IL-2 is best in treating people with melanoma or advanced kidney cancer. The overall response rate for patients with these types of cancer varies across different studies, but is around 5–20%.[116] A small fraction of those who respond are left with no trace of cancer remaining; they are truly cured.

Why IL-2 works for only some types of cancer is not clear. Melanoma, the type of cancer which Taylor had, involves more mutations than most other cancers. So one possible reason why IL-2 helps against melanoma more than most other cancers is that their large number of mutations mark out melanoma cells as being especially different from healthy cells, making them relatively easy for the immune system to detect and react against. Why some patients respond well to treatment with IL-2, but others don't, remains, unfortunately, unknown. It is possible that the treatment works best in people with a level of immune reaction already ongoing against their tumour, there to be boosted by the treatment.

Altogether, this band of pioneers, from Lindenmann and Isaacs to Gutterman and Rosenberg, discovered the existence, and then the power, of cytokines. They seeded an enormous scientific endeavour – cancer immunotherapy – which now has hundreds of branches, each studying a different way of boosting our immune response to cancer. A multitude of cancer treatments, with many more on the horizon, are the outcome. We will return to this endeavour later. Now we will turn to an altogether different therapeutic revolution that came from our understanding of cytokines, not in the fight against cancer but for treating autoimmune diseases, not by boosting immunity, but by stopping it. Enter the *anti*-cytokine.

4 A Multibillion-Dollar Blockbuster

New medicines often begin with big ideas, and Sir Marc Feldmann had one. Born in Poland in December 1944, his family moved to France immediately after the war, and then to Australia when Feldmann was eight. 'Immigrants have a very strong motivation to work hard and succeed,' he thinks, having gained his own work ethic from his father who worked long hours as an accountant while studying at night.[1] At medical school in Melbourne, Feldmann was bored by rote learning of human anatomy, but thrilled by the uncertainties and emerging ideas he found in scientific papers. It was while doing research for his PhD at the Walter and Eliza Hall Institute in Australia – 'fuelled by coffee and the music of the Rolling Stones'[2] – that he took the first real step on the journey towards his great contribution to immunology, ultimately providing relief from pain to millions of people and spawning what is now a multibillion-dollar industry.

It began with a feeling of dissatisfaction. It had recently been established that an immune response involved many different cells – emphasised by Steinman's discovery of dendritic cells (Chapter Two) – and, looking down his microscope, Feldmann could see that immune cells were dynamic and moved about.[3] By contrast, the research he was doing on these isolated immune cells seemed too reductionist, too far removed from what really happens in the body. 'Concepts generated in one precise circumstance often do not extrapolate to complex and non-reductionist reality,' he later wrote.[4] Of course, all scientific experiments are reductionist in some way; it would be impossible to conclude much – if anything – from their outcomes without isolating to some degree the particular aspect of the whole whose effects we wish to study. But Feldmann's point was that he wanted to know what's happening in the body, across the system, not just what's going on inside a single type of immune cell. His thoughts turned to the way that different immune cells talk to one another.

To study this, he set up a flask containing two glass tubes, one inside the other, with a porous membrane at the ends of both tubes, and filled the flask with culture broth. The broth could flow freely through the membrane but larger particles, such as cells, could not. With this set-up he could put different types of immune cells in the inner and outer tubes and keep them separate, while allowing them to bathe in the same culture broth. He set up several of these flasks and by comparing what happened in them with ones without separate tubes so that cells moved around and interacted with each other freely, he could assess which kinds of immune reactions required direct contact between cells and which could be triggered by secretions from cells into the liquid. A handful of other scientists were doing similar experiments around the globe, and even though they had almost no understanding about what was in the liquid specifically, they were essentially all studying the effects of cytokines. In 2016, I asked Feldmann what he learnt from these early experiments, and he replied, with a chuckle: 'We discovered that life is complicated.'[5]

Feldmann was one of the pioneers present at the first cytokine workshop in 1976, among around forty scientists who gathered in a hotel near the US National Institutes of Health with the aim of establishing a coherent picture of what cytokines did. It was an almost hopeless task at first, because there was no way of isolating different cytokines and therefore establishing if the various effects on each type of immune cell were caused by one or several of them.[6] It was only after the cytokine genes had been isolated, allowing the different cytokine proteins to be produced individually, that the effects of each could be studied systematically. This showed that each cytokine had multiple diverse activities, which was a controversial idea at first because it was generally thought at the time that each kind of protein in the body did only one job.[7] Many acronyms used to name cytokines had to be abandoned because several of them, it turned out, applied to the same molecule. Eventually, with the tools in hand to properly dissect the cytokine world, the excitement of a gold rush ensued, electrifying (or intoxicating, depending how you view such things) some of the scientists involved with the prospect of money and fame.

Science's supposed moral purity took a hit in October 1984 at the fourth cytokine workshop. At an exclusive resort in the Bavarian Alps, Philip Auron from Charles Dinarello's lab at Massachusetts Institute of Technology (MIT) announced that his team had isolated the gene for one of the forms of the cytokine named IL-1.[8] One scientist who was in the audience vividly remembers the excitement,[9] and another recalls it as 'the big moment of the workshop'.[10] At the start of his speech, the session chair announced that photography was strictly prohibited, a condition that Auron made before he agreed to present his data. During his talk, Auron flashed up the genetic sequence for IL-1, briefly.[11] As soon as Auron's talk ended, someone dashed to the audience microphone and shouted 'This is not IL-1.'[12]

The heckler was Christopher Henney, who, in 1981, had co-founded the biotech company Immunex, based in Seattle.[13] Henney was approaching forty years old when he started Immunex with colleague Steven Gillis, both then at Seattle's Fred Hutchinson Cancer Research Center. 'I couldn't see myself doing the same thing for twenty-five more years,' Henney says. 'Some guys get their hair permed, put gold chains around their neck, and go chase girls. I decided to start a company.'[14] At the audience microphone, Henney announced that his company had isolated the IL-1 gene, and that the sequence Auron just showed wasn't it.[15] Auron asked Henney to show what he thought was the correct gene for IL-1. But Henney refused and went back to his seat.

A summary of the workshop, published shortly afterwards, noted that 'this lack of etiquette was most surprising … especially since Christopher Henney had a long and distinguished university career prior to joining Immunex'.[16] Soon after the event, Immunex published the sequences of two forms of the IL-1 gene, alpha and beta, in the journal *Nature*. One of these was in fact identical to the gene the MIT researchers had found and announced at the workshop.[17] The MIT team published a letter in *Nature* stating that this proved they were right all along and that Immunex had had no reason to cause such outrage at the cytokine workshop.[18] But more than pride was at stake. A small biotech company, Cistron, were working with the MIT team. They and Immunex had both filed patent applications around the IL-1 genes. Digging into who did what and when to resolve the patent squabble, it emerged this was not a simple case of normal scientific competition.

Cistron alleged that Immunex had cheated. They claimed that Immunex co-founder Gillis gained information about one of the IL-1 genes when he was sent the MIT team's paper by *Nature* for peer review, supposedly a confidential process.[19] *Nature* rejected this paper on account of the peer reviews they received for it. Crucially, the patent application submitted by Immunex included errors in the gene sequence identical to errors found in the draft paper from MIT – something which is highly unlikely to have happened by chance. Cistron alleged that this proved Immunex submitted their patent for IL-1 using the genetic data that they had been sent for peer review. Immunex countered that it was a simple clerical error. And their lawyers built a case claiming that in fact, there were no hard rules dictating confidentiality during the peer review of scientific papers anyway.[20]

It took twelve years for the situation to be resolved and the final settlement involved Immunex paying $21 million to Cistron.[21] Reportedly, both Henney and Gillis helped with the payment personally.[22] By that time, Cistron had gone through bankruptcy, and the value of the disputed patents had anyway become limited as high doses of IL-1 proved to be toxic.[23] Immunex, on the other hand, had by this time discovered and studied a long list of genes that are important in the immune system and in 2002 the company was purchased by another biotech company, Amgen, for $16 billion.[24] Henney and Gillis moved on to be directors of several other biotech companies.

A year before the IL-1 debacle kicked off in the Alps – while Immunex and Dinarello's team at MIT were racing to isolate cytokine genes – Feldmann was on holiday in a small town dominated by a fifteenth-century castle ruin on the Costa Brava coast of Spain. Relaxed and away from the hubbub, he had an epiphany. He worries that long holidays are often shunned in labs nowadays: 'Holidays,' he later wrote, 'provide not only an opportunity to enjoy family, friends, and the splendour of our planet, but also time to think creatively and strategically.'[25] His big idea, which he published later in the *Lancet*, was about the origin of autoimmune disease.[26]

He mused whether immune cells might activate each other through their cytokine secretions to such an extent that the activation becomes self-perpetuating, creating a vicious circle that overstimulates the immune system and causes it to harm the body. This was a powerful new idea. Though he had little proof of it being true he put it out there nonetheless, he recalls, in the 'slightly overconfident mode of the young'.[27] Today, it would be almost impossible to publish an idea like this without a wealth of supporting data, especially in a journal as prestigious as the *Lancet*, but this was a different age for biological science: there were fewer scientists, less competition for space in journals, and editors were probably more open to papers about an idea than they are today. At any rate, ideas can sometimes move us forward even in the absence of evidence, and the most important implication of this one, from a medical point of view at least, was that blocking a cytokine might stop immune cells from driving each other on and thus prevent autoimmune disease.

Feldmann decided to focus on one autoimmune disease in particular, rheumatoid arthritis, a long-term inflammation in joints which causes pain, stiffness and sometimes disability. It affects around one in a hundred people in every country.[28] We don't understand precisely how the problem starts, which probably varies in different people, but the symptoms come about because immune cells accumulate in joints and, over time, cause the destruction of cartilage and bone. To a small extent, rheumatoid arthritis runs in families and forty-six genes have been linked to the disease.[29] But if a person with rheumatoid arthritis has an identical twin (who shares the exact same set of genes), there's still only a one in five chance that their twin will also develop the disease. That's because there are many non-genetic factors involved, which we don't understand very well. Drinking lots of coffee (defined as four or more cups a day in one study) correlated with a slight increase in risk, for example, in one analysis.[30] This connection is not entirely clear, however, because different studies have come to different conclusions,[31] but even if taken as fact, it's hard to disentangle whether or not this is a direct effect of drinking lots of coffee or if drinking lots of coffee is simply indicative of some other cause. At the time Feldmann set his mind to working on rheumatoid arthritis, the feeling among experts was that this was a very complicated disease, with many factors involved, so no simple treatment, certainly not a drug which targeted one particular molecule, would be likely to help.

After his PhD, Feldmann had moved to London, in part because 'there was more money for research than in Australia'.[32] Here the clinician Sir Ravinder 'Tiny' Maini helped Feldmann focus on rheumatoid arthritis. Born in Ludhiana, India in 1937, Maini had moved to Uganda in 1942, where his father became a minister in the British Ugandan government, and then to the UK in 1955.[33] He was recommended to Feldmann as a physician open to new ideas. Two days after they first spoke on the phone, Maini was in Feldmann's office in London, and a long-lasting friendship began. 'It was a meeting of minds,' Maini recalls.[34] Although friendship isn't strictly required for a successful collaboration, Feldmann thinks the kind of trust that friendship brings is essential.[35] They each brought to the table different experiences and backgrounds – Feldmann was better known as an immunologist while Maini had the relevant clinical expertise – but they had enough in common to be able to communicate with the same jargon. Neither led the other; they became two halves of a team. 'A union,' as Maini puts it.[36]

Choosing to tackle rheumatoid arthritis as opposed to other autoimmune diseases was important because the relevant human tissue was accessible to study. Maini could easily provide needle samples from patients' joints, whereas the relevant tissues required for the study of other autoimmune diseases were, and still are, very hard to obtain – the brain in multiple sclerosis or the pancreas in diabetes, for example. They decided that their first goal should be to find out which of all the cytokines were made by immune cells accumulated in the inflamed joints of arthritis patients.[37] Studying cells and fluid isolated from patients' joints is what set Feldmann and Maini apart from most other researchers, and it's what got them on the right path towards finding a way to tackle the disease. They discovered that many cytokines were present but that one – with the unwieldy name 'tumour necrosis factor alpha' (often abbreviated to TNF)[38] – was especially abundant.[39]

TNF had been identified in 1975 as a factor released from immune cells that was able to turn tumours black and dead.[40] This immediately led to great interest in the cytokine in the hope that the tumour-killer could be used to treat cancer patients, a hope that was dashed when it became clear that the cytokine itself is quite toxic to the body even at doses too weak to be able to impact a tumour. But every cytokine has a multitude of activities and TNF's ability, at high doses, to kill tumours was not what interested Feldmann and Maini. Rather, they wanted to test what would happen if they blocked the activity of TNF in the inflamed joints of arthritis patients. To do so, they needed an anti-cytokine – which is something that can be produced in the form of an *antibody*.

Antibodies are secreted by the white blood cells known as B cells and are our body's 'magic bullets', a term coined in the 1890s by German Nobel laureate Paul Ehrlich. They are soluble protein molecules that stick to and neutralise all kinds of germs and other potentially dangerous molecules. Each individual B cell produces an antibody with a uniquely shaped tip, the part of the antibody that sticks to its target molecule, called an antigen, which might be, for example, something on the outer coat of a bacteria or virus. However, antibodies are not designed to bind to germs per se. The shape of each antibody's tip is created almost randomly by a process of chopping up and rearranging the genes that create the antibody, a remarkable process in its own right. B cells which happen to have made an antibody that could stick to healthy cells and tissues are killed off (or inactivated) so that the only B cells allowed in the blood-stream are the ones that make antibodies that stick to something not normally found in the body. This is the process we first met in Chapter One, and is how these cells are able to distinguish *self*, components of your body, from *non-self*, anything that's not part of you.

In more detail, every B cell also has a version of its own antibody tethered to its surface (the B cell receptor we also met in Chapter One), so that the cell can tell when there is some-thing in the body that its antibody could lock onto. When a B cell does have the right antibody to lock onto something alien and troublesome, the B cell multiplies so that its useful antibody is produced in bulk, ready to neutralise the intruding molecule or germ. With around 10 billion B cells in the average person's immune system, each of us has the ability to make around 10 billion differently shaped antibodies, each of which is able to recognise something that hasn't been in the body before, ensuring that antibodies can be produced against virtually any structure alien to the body. This is essential if our immune defence is to tackle germs which the body hasn't seen before – even germs which have never even existed. Crucially for Feldmann's and Maini's purposes, it also means that any animal could make antibodies against a protein found in any other animal. Therefore a mouse immunised with the cytokine TNF could produce antibodies that would lock onto the human cytokine and stop it working – an anti-cytokine.

Precisely such an antibody was made by scientist Jan Vilček at the New York University Medical School. Vilček was born in Czechoslovakia in 1933, to Jewish parents who were proud of their ancestry but not religious. In 1942, his family were granted exemptions from the normal degrading restrictions placed on Jews so that, for example, they were not required to wear yellow Stars of David and could keep their jobs.[41] Such exemptions involved hefty administrative fees, possibly bribes, and conversion to Christianity, but officially, they were given on the basis that the country needed people in certain jobs to continue; Vilček's mother was an ophthalmologist and his father worked in the coal-mining industry, presumably both important enough.[42] At age eight, Vilček was sent for protection to live in an orphanage run by Catholic nuns.[43] After German troops suppressed an insurrection by the Slovak resistance in 1944, Vilček's parents were worried that a new hard-line government wouldn't recognise their exemptions from restrictions on Jews. Vilček left with his mother to shelter for many months among a handful of remote farming peasants, and later in an isolated village.[44]

After the war, with his family reunited, Vilček went to medical school in communist Czechoslovakia, where 'a general atmosphere of fear and suspicion permeated life'.[45] At this time, the term 'gene' was practically banned because Stalin favoured a direct inheritance of traits acquired in one lifespan, and many scientists who disagreed were imprisoned.[46] It was in 1957 during his time as a student that he was inspired to study cytokines after hearing a talk by Alick Isaacs, the co-discoverer of interferon, who was on a visit to Czechoslovakia.[47] As Vilček spoke English very well, he was selected as Isaacs' guide and so got to know him a little.[48] Later, this proved to be an important personal connection.

After medical school, Vilček joined a research centre dedicated to viruses, and in 1960 published a paper in *Nature*, providing evidence for the existence of interferon.[49] The director of the research institute had wanted Vilček to publish in the local journal *Acta Virologica*, but – importantly, it would soon turn out – Vilček did not take the advice.

In 1964, Vilček and his wife Marica, an art historian, defected from Czechoslovakia.[50] Because he had published in *Nature*, not

Acta Virologica, he had three job offers before he even made it out of Europe to his new home in the USA.[51] He picked New York University Medical School, where he has since stayed for his entire career. Vilček later learned that another reason he was offered a faculty post without an interview was that they had received a recommendation letter from Isaacs, whom Vilček had acted as a guide for seven years earlier.

Vilček's life story is one of amazing achievement following an early life of adversity. With the royalties Vilček received from making an anti-TNF antibody, he set up the Vilček Foundation, which champions the contributions of immigrants to life in the USA. In 2005, he had received so much money from making this antibody that he could donate to New York University Medical School the largest gift any New York healthcare institute ever received: $105 million.[52] This bought new professorships, lab refurbishments, a student dormitory, research fellowships, studentships and more. Occasionally – very occasionally – a career in academia can be financially lucrative. Vilček laughs at the thought of it all: 'Getting rich was not really something that was ever a goal of mine. To be honest, I'm still a little embarrassed by it.'[53]

To make the antibody, Vilček first had to obtain a sample of the human cytokine, TNF, to be injected into mice. In late 1985, the company Genentech had isolated the gene for TNF and obtained significant quantities of the protein by expressing the gene in bacteria. Vilček was able to obtain samples of it in 1988 because he was collaborating with them on another project. To make the antibody using mice, he then followed a method that had been worked out by César Milstein and Georges Köhler in Cambridge in 1975, an enormously important method for which Milstein and Köhler won a Nobel Prize in 1984.[54] First, Vilček immunised mice with Genentech's TNF protein and, after a few days, isolated B cells from their spleens, knowing that many of these B cells would be producing antibodies against TNF. Outside the animal's body, B cells can't survive for long, a few weeks at best when incubated in cell-culture broth, but Vilček used a trick to keep them alive – the Nobel Prize-winning trick that Milstein and Köhler had hit upon[55] – which was to fuse the B cells with myeloma tumour cells and create new cells, called hybridomas, which retain the growth traits of a tumour with the antibody-producing abilities of the original B cell. In effect, this creates immortal versions of the mouse B cells. Vilček then isolated each single hybridoma cell, separating them by pipetting a minuscule amount of the liquid suspension that contains the cells into each of the many indentations in a hand-held rectangular plastic dish. Then, the antibody produced by each one could be tested for its ability to block the activity of TNF. The cell found to produce the appropriate antibody was then cultured to produce an almost limitless supply of anti-TNF antibody.

This type of antibody is called a *monoclonal* antibody as it derives from a single B cell. The process can be used to create a protein shaped to lock onto any molecule of our choosing. As well as being used as medicines, antibodies are used in all kinds of scientific experiments, to mark out particular cells, block the activity of something or switch on the activity of something else, check the level at which something is being secreted, and so on. 'No single class of reagents stirs our creativity, or propels our goals, our successes, even our dreams, with as much excitement as do monoclonal antibodies,' as one expert has said.[56]

Vilček had a long-standing agreement with the then fledgling company Centocor to develop commercial applications from the antibodies made in his lab.[57] In return the company paid for some of Vilček's lab costs,[58] including the salary of the postdoctoral researcher, Junming 'Jimmy' Le,[59] who helped make the anti-TNF antibody.[60] It was, however, elsewhere in New York where it became apparent that an antibody against TNF might be medically important. Bruce Beutler, the Nobel laureate we met in Chapter One, who helped discover that toll-like receptors lock onto bacteria, worked, earlier in his career, with Anthony Cerami at the Rockefeller University Hospital, and discovered the mouse version of TNF. In 1985, he found that TNF was one of the cytokines produced in mice during sepsis, a disease caused by an immune response going into overdrive, usually because of a bacterial infection.[61] Importantly, Beutler and Cerami found that blocking TNF could protect mice from the symptoms of sepsis.

In humans, sepsis (called septic shock when symptoms include a drop in blood pressure) can kill patients in a matter of hours and, with many cases not being easily treated with antibiotics, the illness accounts for up to tens of billions of dollars in hospital care in the US.[62] Centocor had already highlighted treatments for sepsis as an important focus for the company.[63] Now, inspired by Beutler's and Cerami's work, they wanted to try treating sepsis in humans by blocking TNF.

They couldn't use Vilček's anti-TNF antibody in people straight away. Since the antibody was made in a mouse, it had to be modified to more closely match antibodies naturally made in humans. Otherwise, the antibody itself would be seen as something alien in the human body and could trigger an immune reaction. To avoid this, segments of genes for the mouse antibody were combined with human genes to create a new half-mouse half-human antibody.[64] Actually about 34% mouse, 66% human, its front end was kept in its mouse form to lock onto the TNF cytokine, and the back end made human. A chimera, like a lion with a head of a goat, made real on a molecular scale.

In 1991, Centocor tested the chimeric antibody in patients with sepsis.[65] Although there were no adverse side effects noted, there was no clear therapeutic benefit. What worked in mice didn't work as well in humans; a common theme in medical research. It looked like the anti-TNF antibody might end up as a scientific tool – perhaps part of a diagnostic test for cytokine levels in blood – but not an actual medicine.[66] Then, in early 1991, Feldmann visited Centocor to present his case for trying the antibody in patients with rheumatoid arthritis.

By that time, Feldmann had some evidence to support the idea that TNF was important in rheumatoid arthritis and that blocking its activity might help. Maini's team had found that the cytokine was present in the right place at the right time for being involved in the symptoms of the disease. In Feldmann's team, Fionula Brennan (who sadly died young from breast cancer in

2012) had looked at what happened when an anti-TNF antibody was added to cells taken from the diseased joint of patients. The result was a eureka moment. Brennan discovered that when TNF was blocked, other cytokines stopped being produced by the cells.[67] She repeated the experiment seven times to be absolutely sure.[68] This implied that TNF was at the top of a cascade of events, or the hub of a network, which led to the other inflammatory cytokines being produced. These results flew in the face of the scientific consensus at the time, that no single molecule could be responsible for something as complex as inflammation in rheumatoid arthritis. Also, most scientists thought that the cytokine system involved lots of redundancy, so that if you blocked one component, it wouldn't make much difference to inflammation overall, because other cytokines would continue to work. The dogma was wrong, Feldmann argued, and blocking this one cytokine, TNF, might stop this autoimmune disease.

Another of Feldmann's team, Richard Williams, tested the idea in mice. The symptoms, though not the underlying cause, of human arthritis were recreated in mice by immunising the animals with collagen so that an immune reaction develops against this protein, a major component of cartilage, which in turn results in swelling of the animal's joints. The afflicted mice were then given an injection of anti-TNF antibody and, at high doses, the inflammation was reduced and cartilage at the animal's joints was spared from damage.[69] This showed that mice could be relieved of the symptoms of arthritis by an injection of anti-TNF antibody.[70] Still, many at Centocor were sceptical of this working in people – in part because the company's only rheumatologist had his own ideas about what would work in treating rheumatoid arthritis.[71]

Importantly, James 'Jim' Woody, who had done research for his PhD under Feldmann's supervision in London, was now the chief scientific officer at Centocor and he liked the idea. In fact, he took up the position at Centocor knowing that this opportunity was on the horizon, and he helped prepare the way by involving Feldmann in other projects at Centocor, so that when Feldmann pitched his idea for treating rheumatoid arthritis, he was already known in the company as a leading academic. Vilček thinks that without Woody there, supporting his former boss, Centocor wouldn't have tried the anti-TNF antibody in rheumatoid arthritis patients because it was 'such a long-shot'.[72] But Woody *was* there and Centocor agreed to provide enough of the antibody for a small trial. A personal connection is often what's needed to make things happen.

Centocor agreed that Feldmann and Maini could conduct a small trial at Charing Cross Hospital in London, with just ten patients at first, then later another ten, all tested without placebo controls. The reason they didn't use placebo controls was that at this point they thought of this as a scientific experiment, to test if blocking TNF could do anything to help patients, and – though it seems strange in hindsight – they weren't really thinking that the anti-TNF antibody might itself be used as a medicine.[73] Maini recruited patients who had not responded to any other medicine available at the time. Everyone he approached was happy to participate, despite him explaining the risks of an experiment that was possibly dangerous.[74]

Centocor's trial in 1991 in sepsis patients had showed that the antibody was at least broadly safe, but Feldmann and Maini were nevertheless cautious and each infusion was started slowly. For the first patient, treated on 28 April 1992, they had a nurse spend the night in the patient's room. They needn't have worried. As the antibody was infused, many patients said they felt better immediately. 'It was a very thrilling time,' Feldmann recalls, 'all the patients we treated improved dramatically.'[75] Graphs, bar charts and statistical analysis recorded the results – the reduction in swelling and tenderness in patients' joints was formally significant after two weeks[76] – but video footage of patient number eight said it all.

Before treatment, patient eight walks very slowly up and then down some stairs, one step at a time, holding onto the handrail, clearly in pain. Four weeks after treatment, she runs down the same steps as fast as anyone could. At the bottom, she throws her arms up in the air and exclaims ta da! The happiness on her face is a reminder of what this story is all about.

Vilček remembers seeing the video in Centocor's offices soon after it was recorded.[77] Even though this small trial didn't have any control patients, it was clear to him that such a profound change in the patient's well-being could not have been brought about by the placebo effect, a consequence of the brain's anticipation of a positive response. But there was still a huge question to be answered: how long would the benefit last? Everyone involved – patients, clinicians and scientists – knew that the next few months were vitally important. All the patients returned to their normal lives, health improved. A dentist – patient number three – was able to play golf just two weeks after his treatment and later returned to work.[78] But unfortunately the benefits were short-lived. Everyone relapsed.

Evidently the antibody wasn't a cure; but it could relieve symptoms. This meant that the next logical step was to test the benefits of blocking the cytokine repeatedly. Feldmann and Maini obtained ethical permission to re-treat some of the patients, and again, all those tested improved. Still, the results were anecdotal, with no controls; just twenty patients had received the first dose, and after relapse, eight had been re-treated. As a scientific experiment, it was informative, but a medical advance requires a proper clinical trial, randomised and double-blind, where neither clinicians nor patients know who is getting the new treatment and who isn't.

The results of the first formal trial were unequivocal; anti-TNF antibody improved the health of rheumatoid arthritis patients. Detailed analysis of what was happening in patients' blood revealed that the antibody was working just as Feldmann and Maini predicted: blocking this one cytokine reduced the production of other inflammatory cytokines, and biopsies showed that fewer immune cells were entering the diseased joints. Feldmann thinks such detailed analysis of what was happening in patients – taking biopsies and analysing 400 ml of blood from everyone – was done because the trial was carried out by academics rather than a company. Most clinical programmes, Feldmann believes, don't spent the time and money on analysing what happens in patients to such extent, and this is a great loss.[79]

That's not to say it all ran smoothly in the hands of the academics. At one point, a freezer defrosted and crucial samples were destroyed; 'It was physically painful to think of the major scientific opportunities lost,' Feldmann recalls.[80] For the next clinical trial – a full-scale comparison of anti-TNF antibody with existing treatments, a so-called phase III trial – Centocor was keen to get their antibody approved as a medicine as fast as possible, so fewer samples were taken and, Feldmann says, the emphasis on detailed analysis was lost.

The phase III trials proved that anti-TNF antibody was an effective therapy and was better than other treatments available at the time. Experiments in mice revealed that its benefits were enhanced when combined with other drugs that helped dampen immune responses. This led directly to what is commonly prescribed to patients today: anti-TNF antibody usually taken with another drug, methotrexate, which has many effects in the body including, it turns out, dampening T cell immune responses. This was an early example of medicines being used in combination to treat a disease, something that's far more common today. Poly-pharmacy, Feldmann calls it.[81]

Centocor's ambition to treat sepsis was never fulfilled; the disease remains notoriously difficult to tackle, likely because a storm of inflammation builds up so rapidly in the body that it's especially hard to control. When it became clear that its efforts to treat sepsis weren't working out, Centocor's share price collapsed, down from $50 per share to just $6 within a few months in 1992, and the workforce was reduced from around 1,600 to 400.[82] Anti-TNF therapy came to the company's rescue. Centocor's human-mouse chimeric anti-TNF antibody was marketed as the drug Remicade.[83] And as a result, in 1999, the company was bought by Johnson & Johnson for $4.9 billion.[84] Vilček recalls that the price seemed high at the time – initially, sales of Remicade were slow because it was such a radical new kind of medicine and physicians did not to rush to use it – but later, $4.9 billion looked like a bargain.[85]

Feldmann regrets that the blocking of TNF became another British invention commercialised in the US. He had approached

UK companies, but they weren't interested; only the charismatic leaders at Centocor took the chance.[86] Many of Centocor's rivals in the US, including Abbott, Roche and Immunex, went on to develop other drugs that also block TNF. Beutler helped create one of the alternatives: a soluble protein version of the cytokine's natural receptor. Effectively this acts as a decoy receptor to prevent the cytokine's engagement with its real receptor on immune cells. Clinical development of Beutler's drug, led by Immunex, began two years behind Centocor's but raced ahead to such an extent that in November 1998 it became, in fact, the first anti-TNF medicine approved to treat rheumatoid arthritis in the US, marketed as Enbrel.[87] Other US companies made alternative versions of anti-TNF antibody, including a fully human version available from 2002.[88] All of these medicines have been very successful. By convention within the pharmaceutical industry and according to, for example, the European Commission, a drug is considered a blockbuster if sales reach $1 billion, which makes Centocor's anti-TNF antibody a blockbuster many times over. UK-based companies largely missed out on a therapy which, in 2012 alone, made $9.3 billion. This could be viewed as important or trivial, but what is undoubtedly important is that because of drugs that block TNF, far fewer people with rheumatoid arthritis are forced to use a wheelchair.

If rheumatoid arthritis were the only condition that could be treated by blocking TNF, then this therapy would still be a blockbuster, but it proved to be of even wider use. Blocking this cytokine helps stop inflammation in many situations where it is a problem: in the digestive system, as happens in Crohn's disease and colitis; in skin, as happens in psoriasis; and in the joints of the spine in ankylosing spondylitis. Across the globe, Centocor's anti-TNF antibody has been used to treat at least 1.8 million people.[89]

This success did not come about in any easy or linear way; it took a multitude of small steps for Centocor to make the part-mouse part-human antibody, tested by Feldmann and Maini, based on the antibody Vilček made first. Triumph came from imagination and hard work, but also from a web of coincidences, chance events and serendipities. Reflecting on this in his memoir, Vilček quotes E. B. White, the author of children's books *Stuart Little* and *Charlotte's Web*: 'No one should come to New York to live unless he is willing to be lucky.'[90]

All those who directly contributed to the discovery of anti-TNF therapy deserve celebration. Many have, rightly so, won prestigious scientific prizes. In 2013, Vilček received the US National Medal of Technology and Innovation from President Barack Obama, who summarised Vilček's life in a few moving sentences. Vilček wished his parents had still been alive to hear them.[91] In 2003, Feldmann and Maini won the prestigious Albert Lasker Award for Clinical Medical Research and, in 2014, the Canada Gairdner International Award. But those who pursued other cytokines, in other illnesses, were doing important science too. After all, it might well have turned out that blocking another cytokine would help patients with rheumatoid arthritis – in fact, we now know that blocking IL-6 can also help – and it might have worked out that blocking TNF would help treat sepsis, for example. It takes a community to explore all the possibilities.

Indeed, unlike vaccination – discovered long before anybody had any detailed understanding of how it worked – anti-TNF therapy emerged directly from understanding the molecules and cells that make up our immune system; knowledge that was generated by thousands of scientists. We tell the stories of individuals – and perhaps ego propels scientists into action – but no scientist is an island. At some level, this therapy was achieved by a collective scientific mission to understand immunity. Maini is especially proud of this fact; that his work helped show how the detailed molecular science of immunology can be harnessed for medicine.[92]

★

Science doesn't have any endings; discoveries are made, new treatments found, but everything leads to something else. The discovery of anti-TNF therapy was a watershed moment because it introduced a new way to combat disease – manipulating the immune system rather than directly fighting germs with, say, antibiotics, in a way that's very different to vaccination. Feldmann's next big idea was to wonder how many other illnesses might be tackled with drugs that block cytokines. Although we do not yet know the extent to which asthma, diabetes, coughs, the common cold and strokes can be tackled by manipulating cytokine levels, these and many other conditions and diseases are all potential targets. Pharmaceutical companies and many academic research labs are on the case, betting that the success of blocking TNF is not a one-off fluke but the dawn of something even bigger.

That's not to say blocking TNF is a perfect medicine; far from it. There are at least three significant problems with anti-TNF therapy. First, blocking this part of the immune system inevitably weakens our defence against infections.[93] Major problems are rare, but for people with latent tuberculosis – showing no sign of illness because their immune system normally controls the infection – there is an increased risk of the disease reactivating when their immune system is compromised by anti-TNF medicine.[94]

A second problem with anti-TNF therapy is that a significant fraction of patients do not benefit from it: as many as four in ten rheumatoid arthritis patients show little improvement.[95] Combinations of drugs can improve response rates, but unfortunately, we currently have no way of knowing in advance who will respond and who won't. Standard clinical practice is to proceed by trial and error: patients are simply given one of the drugs that block TNF and if no significant improvement is experienced in three months or so, they are switched to a different type of anti-TNF therapy or something else entirely. Maini thinks that one factor which might be important in determining whether or not a patient responds to anti-TNF therapy is how long ago the problem began.[96] If the inflammation in a person's joints has already lasted a very long time, Maini thinks, it probably becomes – in some sense that it is not well understood – more complex and harder to control. Perhaps this is also one reason why some patients respond well to anti-TNF therapy at first, but the drug loses its potency over time.[97] The third problem is that blocking TNF is an effective treatment, but not a cure.[98] The quest for a cure continues.

Feldmann's and Maini's research also had far-reaching consequences because of the type of medicine they used – an antibody. At the time, the potential for antibodies to be used as medicines was not widely recognised because they were – and still are – very expensive to produce. The B cell hybridomas that produce antibodies have to be grown in a broth that contains something like fifty different ingredients. Even in optimal conditions, stirred just right in a bioreactor, each cell produces only a minuscule amount of antibody, which then has to be purified to a standard that's safe for use as a medicine. A few companies, Centocor included, were built on the premise that antibodies could make money, but even these companies thought that they were most likely to obtain regulatory approval for their use in diagnostic blood tests far more easily than for actual therapies. Centocor's first antibody product was a test for the hepatitis B virus, for example.[99] The therapeutic and commercial success of the anti-TNF antibody showed everyone the true scale of antibodies' potential as medicines.

Anti-TNF antibody was not actually the first antibody approved as a medicine; that was an antibody sold as Orthoclone, approved in 1985, designed to lock onto the white blood cells known as T cells and eliminate or deactivate them. It was hoped that this antibody could stop an immune reaction from developing in transplant patients, which might otherwise cause a transplanted organ to be rejected. This antibody was approved for use in kidney, heart and liver transplant patients but is no longer used today. The antibody didn't work well and the side effects were serious; some people developed a potentially life-threatening condition, likely on account of the treatment sometimes triggering T cells to release high levels of cytokines. Ever since Milstein and Köhler learnt how to make antibodies à la carte, it seemed that there had to be a role for antibodies in medicine.[100] But in practice, for the nearly two decades it took for anti-TNF to be developed, the pursuit felt like chasing a rainbow.

One of the most important antibodies developed subsequently is rituximab. Instead of blocking a cytokine, this antibody directly targets immune cells, specifically B cells. When it locks onto a protein molecule on the surface of a B cell, that particular B cell is destroyed in one of three ways. First, the antibody itself can cause the B cell to self-destruct. Billions of our body's cells die this way every day to allow for a healthy turnover of cells in the body; rituximab can simply trigger this same programme of cell death. A second way that the antibody kills B cells is that, while its front end is tethered to a B cell, its back end attracts factors in the blood which then kill the B cell. Alternatively, its back end can be recognised by the immune system's Natural Killer cells, which flatten up against the B cell and kill it. Again, these last two processes happen as part of our normal immune defence; antibodies usually lock onto germs or infected cells, things which warrant attack in this way. Rituximab essentially causes a person's own B cells to be detected by the immune system as something to be eliminated.

The loss of B cells in the body that results from this antibody can in turn dampen inflammation in a patient's joints, so rituximab is prescribed as an alternative medicine for those rheumatoid arthritis patients who do not benefit from anti-TNF therapy.[101] However, it was first approved not for the relief of rheumatoid arthritis but in 1997 to treat cancer. It has since been used by over 750,000 cancer patients. At a glance, it seems quite unlikely that a cancer drug could help with rheumatoid arthritis; these maladies have little in common. But an antibody which kills B cells is useful for the types of cancer – chronic lymphocytic leukaemia and non-Hodgkin lymphoma – where it is a B cell that has lost control and become malignant. In fact, this antibody is so important that it features on a list of the world's most essential medicines, produced by the World Health Organization, each selected because 'their potential health impact is remarkable'.[102]

Our detailed knowledge of the way in which antibodies kill has led to improvements in their design: for example, antibodies can be produced with subtly different structures that make them more effective at triggering attack by Natural Killer cells. It has also led to the discovery that some people have a genetic variation which makes their Natural Killer cells less efficient at killing cells coated with antibody. In lymphoma patients, there is evidence that this genetic variation correlates with a less successful response to rituximab (but the issue is controversial as this has been found in some, but not all, studies).[103]

Not all antibodies which lock onto B cells are equally effective at causing them to die and so an important scientific frontier is to understand why rituximab is so effective. In my own lab, we have used state-of-the-art laser-based microscopes – which cost around half a million pounds each – to make videos of the process by which rituximab binds to a cancer cell and then causes immune cells to attack.[104] We discovered that rituximab doesn't coat a cancerous cell uniformly but tends to gather on one side of the cell, drawing some proteins towards and into the area where the antibody has accumulated and sending others to the opposite side of the cell. In effect, this antibody creates a front and back to the roughly spherical cancer cell, or in the jargon, causes the cancer cell to become polarised. We found that cancer cells which gained this polarity were, for reasons we don't understand, more easily killed.

Just by watching what happens with the use of a microscope, we can infer that rituximab's effectiveness as a drug is, in part, attributable to this ability to change the structure of the cancer cell, making it especially susceptible to being killed by immune cells. This implies that when making new antibody-based medicines to kill off a particular type of cell, it would be useful to screen not just for those which stick to the right type of cell but those which trigger equivalent changes in the target cell's structure as well. This is, however, at the edge of knowledge – a hypothesis rather than a fact – and it's hard to know for sure how important this is because we can only watch such activity in a lab culture dish. It is alas impossible to watch whether or not antibodies trigger these events inside patients. This was, after all, Feldmann's point at the outset: we need to know what happens inside the body, where the entire system is at work, not just in isolated cells in a lab dish. The lens, used in a micro-scope or a telescope, has opened up all kinds of new worlds, in space, in ponds and in us. New technologies that improve how we see, especially how we see *inside* the human body, will, for a long time to come, play an ever-increasing role in medical research.

The success of the anti-TNF antibody, and rituximab, began a fashion for seeking more antibody-based medicines, but in 2006 momentum was lost. Controversy erupted when a clinical trial testing a different antibody drug, named TGN1412, went horribly wrong. The trial – implemented by a small company that soon became insolvent – used an antibody that was designed to activate the T cell without the usual need for dendritic cells, the alarm cells that Steinman discovered, to detect danger first. The thinking was that these T cells should then attack cancer cells more readily.

In animals, the drug didn't cause any problems, so it was tested in people – thankfully at a low dose. All six patients given the drug suffered heart, liver and kidney failure and 'many months of hell', as one victim put it when interviewed later by the BBC.[105] As we've seen already, drugs often behave differently in people than they do in mice. In the patients, the drug acti-vated T cells to such a high degree that they began to attack the body's healthy cells and tissue. The overly active immune cells also released cytokines at such high levels that they became toxic to the body. What happened to the patients in the trial is somewhat like what can happen in sepsis, an overreaction of the immune system caused by an acute bacterial infection. The patients all developed fevers, one developed pneumonia, their blood circulation began to fail and their fingers and toes went black. Thankfully, nobody died, but the clinical trial was a tragedy.

Many scientists profess to know – in hindsight, of course – that switching on a vigorous immune response, circumventing the body's normal checks and balances, is bound to be a bad idea. An official investigation found that the problems were, however, due to an 'unpredicted biological action of the drug in humans'.[106] Whether or not the disaster was predictable, the fallout was huge and included major changes in how human trials are now approved. For example, a group of patients should never be given a new drug at the same time: an interval between them allows for side effects to become apparent. In this case, patients developed an inflammatory response which could have been picked up within ninety minutes.[107] So waiting even this short amount of time could have spared subsequent patients the same trauma. The important lesson for science was finding out – all too dramatically – that tinkering with our immune system is like trying to harness nuclear power: there is great potential but a mistake can be catastrophic.

Ultimately, the discovery of anti-TNF therapy showed us that a detailed knowledge of immunity pays off, not only because it reveals a hidden beauty in how the human body works but because this is an area of science that leads to new medicines. Still, the road to each new medicine is not a highway; it is a narrow lane, uncharted on the satnav and full of blind corners. Driving fast is unsafe. We must map more of the immune system to understand how and why its activity varies, to understand the boundaries it operates safely within and, crucially, how it connects with other body systems – which is where we will turn our attention next.

Part Two

The Galaxy Within

5 Fever, Stress and the Power of the Mind

One day in early 1996 I stumbled across a way of helping immune cells fight cancer. Aged twenty-five, I had recently obtained my PhD in physics in Glasgow, and had arrived in a lab in Harvard University to study the immune system. The lab head, Jack Strominger, had, in the 1950s, helped discover how penicillin works and had since then turned his attention to how T cells detect signs of disease in the body, work which made him a candidate for a Nobel Prize.[1] His team was chock-full of driven and talented scientists, about twenty in the main Harvard campus where I worked and another twenty in a second lab he ran a couple of miles away at the Harvard Medical School campus, all of whom knew far more about the immune system than I had gleaned from my education in physics. It felt like I'd been let in through some kind of administrative error.

At the time I arrived, Strominger's lab was focused on under-standing how and to what extent our white blood cells called Natural Killer cells can attack cancer cells. To study this, Natural Killer cells were isolated from blood – blood taken from researchers down the corridor – and mixed with different kinds of cancer cells. The cancer cells had been loaded with a radio-active isotope, so that if they were killed, radioactivity would spill out from the broken-apart cancer cells into the broth that the cells were in. Then by measuring the radioactivity of the liquid broth, we could infer what fraction of cancer cells had been killed by Natural Killer cells. One day, perhaps because of my background in physics rather than biology, I wondered what would happen if we heated the cells a bit. I didn't have any hypothesis to test or any prediction as to what would result, I just wondered. So I briefly heated the cancer cells to around 41°C and found out that they were destroyed much more efficiently.

I never pursued this observation, but a few years later other scientists made a breakthrough by understanding why this happens: heat can induce some types of cancer cells to exhibit at their surface 'stress-inducible proteins', so-called because cells display these proteins when they are in a state of stress. Not stressed in the everyday sense of the word but cells undergo what is called a stress response when they are damaged by, for example, exposure to high temperature, toxins or UV light. Protein molecules become misshaped by heat while UV light can break up a cell's genetic material, and if a cell has these problems it will put up at its surface protein molecules which are not found on healthy cells. These proteins act as a hallmark of cells that are damaged and when Natural Killer cells detect them on a cell, they attack it.[2] My own exploration was not at all important here – the big discovery was made by those who understood the process – but it illustrates how science sometimes moves forward when someone asks a question sideways on. It's why many heads of labs like to hire people from different backgrounds. Having become a lab head myself, I now know it was no mistake that I was let into a top-flight biology lab with a PhD in physics. Professors at Harvard know how to get ahead.

The idea of using heat to treat cancer is far from new. In fact, it can be found in the oldest known description of cancer we have: the Edwin Smith papyrus from around 3,000 years ago.[3] Likely a copy of an ancient Egyptian medical textbook, the Edwin Smith papyrus details how hot blades and sticks could be used to tackle breast cancer. This was probably more to do with trying to burn cancer cells than anything more subtle. Modern experiments show, however, that beyond the burning of diseased cells, heat may indeed help treat some types of cancer. In mice with a type of lung cancer, for example, a feverish temperature lowered the chance of the cancer spreading.[4] And keeping mice in cages warmed to 30°C can boost the numbers of T cells that infiltrate and target a tumour.[5] (That's not to say this proves a direct effect; mice kept in a warm environment tend to be less active, drink more, and so on, and any of these effects could feasibly underlie a boost to their immune response.)

In medicine today, temperatures above 50°C are sometimes used to destroy cancer cells directly, through the application of radio waves, for example. And temperatures akin to a fever are sometimes induced locally or in the whole body to boost the effectiveness of chemical agents given at the same time – a treatment known as hyperthermia therapy.[6] But heat is not used to treat cancer routinely. A reason for this is that the relationship between heat, stress-inducible proteins, inflammation and cancer has turned out to be far more complex than anyone could have known at the time I performed my own heat experiment.

For a start, while our immune system can often suppress or destroy cancer, it can also do the opposite, and there are at least two ways in which cancers can benefit from an immune response, problems that may then be made worse by heat. First, many cancer cells co-opt features of immune cells – by expressing sets of protein molecules used by immune cells – so that they themselves respond to cytokines and other secretions produced during inflammation. This allows cancer cells to hijack the cues immune cells use to multiply and move around the body, so that they too grow, expand and spread. Second, solid tumours sometimes benefit from local inflammation because this can increase the tumour's supply of nutrients and oxygen. In fact, immune cells can be so beneficial to cancer cells that instead of evading an immune attack, some tumours secrete protein molecules specifically to attract immune cells to live inside them.[7] These tumours often secrete hormones to alter the nature of the immune response at the site of the tumour, switching off the immune cells' capability to attack while maintaining a tumour-promoting local inflammation.[8] A tumour that maintains a local inflammation is sometimes thought of as a wound that never heals.

Another complication is that cancer cells which display stress-inducible proteins at their surface – the proteins detected by Natural Killer cells as a hallmark of disease – can sometimes secrete soluble versions of these same proteins into their surroundings. These secretions can act as a decoy, sticking to the receptor proteins on immune cells, blocking them from being able to detect the actual cancer cells.[9] But again, the complete opposite can also happen. Some experiments have found that soluble secretions of stress-inducible proteins from tumours prime Natural Killer cells to be even more alert and even better at attacking tumours.[10] In other words, secretions from tumour cells can, in some circumstances, switch off an immune attack and in other situations amplify the attack. This is at the cutting edge of current knowledge where our understanding becomes fuzzy, which explains why it is so hard to know if, when, and for which types of cancer it could be useful to increase or decrease the production of stress-inducible proteins, either by heat or any other way.

Leaving aside the special case of cancer for now, though, the fact that all warm-blooded animals are able to raise their core body temperature during an infection – which we call a fever – indicates that this ability must provide a hugely important survival advantage, especially as it requires a lot of energy; an increase in body temperature of 1°C requires an increase in the body's metabolism of around 10–12%.[11] Even more wondrous is that cold-blooded animals – reptiles, fish and insects – raise their temperature during an infection too. Unable to change their temperature from within, they do this by moving into a warmer environment. Amazingly, the heat-seeking behaviour of an infected iguana or tuna fish can be reduced with medicines which lower a fever in us, such as aspirin.[12] This means that at least some of the chemical and biological processes causing a reptile or fish to seek a warmer habitat during an infection are similar to those within us during a fever. Even plants might be capable of something akin to a fever, as the temperature of bean-plant leaves can increase during a fungal infection.[13]

For most of human history, a fever was seen to be demonic or supernatural, a problem to be cured. Throughout the eighteenth and nineteenth centuries, people were often said to have died of a fever: yellow fever, scarlet fever, dengue fever, typhoid fever, and so on.[14] Doctors used gruesome methods to try to cure a fever, inducing sweating or vomiting, or blood-letting. Now we know that a fever is part of the body's response to disease, not an illness in itself. Fevers punctuate all our lives; a periodic reminder that so much of how we feel is down to our body's basic physiology.

Raising temperature helps the body fight infections in all kinds of ways, affecting germs directly and increasing the activity of our immune system. Most germs that afflict us have evolved to thrive at normal body temperature. As a result, the replication rate of a virus, for example, decreases 200-fold when the temperature is increased to 40–41 °C.[15] A fever also helps the immune system by increasing the number of immune cells entering the bloodstream from bone marrow, where they are produced. As a result of this, and because heat also causes immune cells to make receptor proteins which direct them to sites of inflammation, a fever increases the flow of immune cells to where they're needed.[16] Once the cells are in the right place, all kinds of immune-cell activity can be boosted by an increase in temperature: macrophages are better at engulfing bacteria; B cells produce more antibodies; dendritic cells, those discovered by Steinman, are better at switching on T cells, and so on. But like everything to do with the immune system, the process can overshoot. Though rarely truly dangerous, a fever can sometimes lead to seizures. Far more common is the sense that your mind and your body are no longer your own.[17]

The mental ache of a fever makes plain the bond between our immune system and our mind. It's a feeling that's hard to express in words, even for Virginia Woolf: 'English, which can express the thoughts of Hamlet and the tragedy of Lear, has no words for the shiver and the headache. It has all grown one way. The merest schoolgirl, when she falls in love, has Shakespeare, Donne, Keats to speak her mind for her; but let a sufferer try to describe a pain in his head to a doctor and language at once runs dry.'[18]

The trigger for the body to raise its temperature – in us and probably all animals – is the detection of telltale signs of germs by the immune system's pattern-recognition receptors. These are the receptors, discussed in Chapter One, whose existence Janeway predicted and which were later discovered in flies, then humans. When these receptors lock onto, for example, the outside coat of bacteria or a virus, an immune response begins and as part of this response, cytokines are secreted. As we discussed in Chapter Three, cytokines call into action different types of immune cells. But cytokines also affect the behaviour of many other types of cells in the body, including neurons. In fact, one of the reasons why blocking cytokines proved to be so effective at treating some rheumatoid arthritis patients is that, as well as stopping the inflammation and thereby increasing the mobility of a patient's joints, blocking cytokines also limits the impact of the inflammation on the nervous system, so that patients often *feel* a lot better quickly.[19]

As well as cytokines, the detection of germs by pattern-recognition receptors also triggers the production of the hormone prostaglandin E2. Prostaglandin E2 can be produced by nearly all types of cells in the body, but during an immune response it is mainly produced by immune cells as well as other cells responding to the cytokines produced by immune cells.[20] The production of cytokines and the hormone prostaglandin E2 is essentially how the immune system warns the brain of danger and triggers a fever.[21] Aspirin reduces a fever by stopping prostaglandin E2 from being made.[22] (You may have come across the hormone prostaglandin E2 as the active ingredient in gels or tablets given to pregnant women to induce labour. Its ability to induce labour isn't directly related to its role in a fever; it's just that every hormone and every cytokine has a multitude of effects in the body, and the ability of prostaglandin E2 to relax muscles can help start contractions of the uterus for birth.)

In a fever, these cytokines and hormones act on a region of the brain called the hypothalamus. In response, the hypothalamus signals for the body to produce another hormone, noradrenaline, which constricts blood vessels in the body's extremities and triggers brown fat cells to burn up energy and produce heat (the specialist job for this type of fat cell), as well as acetylcholine, which acts on muscles to cause shivering, for example, all of which serves to increase the body's temperature. The hypothalamus also controls our feelings of hunger, thirst and sleep, as well as more complex emotions such as seeking closeness with others and our sex drive. Because of this, as well as feeling sleepy and losing appetite, secretions from immune cells affect all sorts of behaviours and emotions. Although this is not very well understood in detail, our immune system undoubtedly shapes our moods and feelings. Some of this might just be a chance outcome of the way in which hormones and cytokines are interconnected, but some of this is likely to have evolved for a reason. There's an advantage in, for example, seeking comfort from others who may care for you when ill. Music, it seems, is not the only food of love; caring affections can be fired up by the chemical reaction of immune cells detecting germs.

Broadly, the immune system and our nervous system are in constant dialogue, each affecting the other through the body's flux of cytokines and hormones. Many hormones affect our immune system, including the sex hormones oestrogen and testosterone, but it is stress hormones that have the greatest impact. We all know what stress is, though it's hard to define. It can be as all-encompassing as a fever or as fleeting as butter-flies in the stomach. What is clear is that stress can have major effects on our health, because of its connection with the immune system. Reducing stress may boost immunity, for example. And our knowledge of the connection between stress, hormones and the immune system has led to one of humankind's greatest ever medical triumphs – as we shall now see.

★

On 1 April 1929, American physician Philip Hench had a routine appointment with one of his patients at the Mayo Clinic, Rochester, Minnesota. This sixty-five-year-old patient happened to mention that the pain he suffered from rheumatoid arthritis lessened while he had jaundice (a yellowing of the skin usually caused by a problem in a person's liver). The patient told Hench that a day after jaundice appeared, he could painlessly walk a mile, something that he couldn't have done before. Hench, a fan of Arthur Conan Doyle's detective Sherlock Holmes,[23] seized on his patient's comments as a clue. He wondered if something in the body, induced when a person had jaundice, could alleviate rheumatoid arthritis. He called it substance X.

Over the next few years Hench came across others with a similar experience and noted that patients with jaundice were often relieved from all kinds of problems, not just rheumatoid arthritis, but hay fever and severe asthma too.[24] He also began to record anecdotes from pregnant women with rheumatoid arthritis who said that they too gained relief from arthritic pain while pregnant. By trial and error, Hench set out to identify substance X. He injected or gave orally liver extracts, diluted bile, even blood, to try to help arthritic patients. Everything failed.

Elsewhere at the Mayo Clinic, biochemist Edward Kendall was on a different mission: to isolate the hormones produced by the adrenal gland – the term 'hormones' having been first used relatively recently, in 1905, by London-based physiologist Ernest Starling, to describe 'the chemical messengers which speeding from cell to cell along the bloodstream, may coordinate the activities and growth of different parts of the body'.[25] At the University of Basel, Polish-born chemist Tadeusz Reichstein independently worked towards this same goal.[26] To give a sense of the effort this required, a tonne of adrenal tissue from cattle, received from a slaughter house, would yield about twenty-five grams of active hormones.[27] Kendall had separated out a number of them, which he simply designated A through to F. One, which Kendall had named compound E and Reichstein called substance Fa, was especially biologically active, based on experiments in animals. The breakthrough happened after Hench and Kendall discussed their seemingly different lines of research in January 1941.[28]

Hench knew nothing about compound E and Kendall knew nothing about rheumatoid arthritis, but chatting with one another over coffee and sharing their separate experiences, an idea hatched.[29] Hench and Kendall decided that it would be worth testing whether or not the adrenal compound E could be substance X. Even if it wasn't, the results would probably be interesting. Hench noted the plan in his notebook but it took almost eight years before enough of compound E was available to test the idea.[30] On 21 September 1948, a twenty-nine-year-old woman from Indiana with debilitating rheumatoid arthritis was treated with compound E, which, in the end, Kendall had obtained from the pharmaceutical company Merck. Two days later she could walk again, and to celebrate, she left the hospital for a three-hour shopping spree.[31]

Luck had played its role. Hench happened to have guessed a dose of the hormone which worked well – a dose higher than most physicians would have thought OK to try – and used crystals of it which happened to be the right size to be dissolved in the body at an appropriate rate.[32] Luck helped in less scientific ways too: when the precious sample of compound E first arrived in the hospital, the glass vial fell to the marble floor, but didn't smash.[33]

When Kendall was invited to meet the patient, she rose from bed and said 'Let me shake your hand.'[34] As a chemist, he seldom met patients and the moment was hugely significant for him, the culmination of eighteen years' work. Hench also understood the magnitude of their success – so much so that he insisted they rename the compound as compound H, and that nothing should ever be discussed about it over the phone, in case someone else would scoop their discovery.[35]

Over the next few months Hench treated other patients. Many had been confined to a wheelchair but were soon on their feet. Hench presented the results for the first time at a meeting primarily for his fellow staff at the Mayo Clinic on 20 April 1949.[36] Rumours had already spread that something big was going to be announced and the room was packed. Probably because he had a speech impediment, Hench was one of the first lecturers at the Mayo Clinic to use slides and other visual aids.[37] On this occasion, he showed a flickering colour film of patients before and after their treatment, at a time when most film and photography was black and white and TV was a novelty. Changes in the patients were remarkable, and the moment was made all the more emotive because many in the audience knew the patients personally. The film sparked resounding applause even before it finished. After showing the film, Hench approached the lectern and received a standing ovation.[38] Kendall spoke next, and emphasised how basic chemistry underlies new medicines.[39] Soon after, in 1950, Hench, Kendall and Reichstein won a Nobel Prize. Never before or since has a Nobel Prize been awarded so rapidly.[40]

We now know that among the hormones produced by our adrenal glands in response to stress, one that is especially significant to the immune system is cortisol.[41] Cortisol works to prepare the human body for stressful situations by helping establish, for example, the body's fight-or-flight response: increasing our blood sugar levels and dilating blood vessels for muscles to prepare the body for immediate action. Importantly, cortisol also quietens the immune system, perhaps to prevent an inflammatory reaction switching on or overshooting when the body is under stress, and perhaps also because an immune reaction isn't of immediate importance in a fight-or-flight situation and energy is best used elsewhere. Overall, cortisol has an incredible impact on the human body, affecting the activity of around one in five of all 23,000 human genes.[42]

Substance X, compound E, substance Fa, compound H or, more precisely, the compound Merck managed to synthesise, was named cortisone (it is very closely related to cortisol; enzymes in the body can change one into the other).[43] And it quickly became the most sought-after drug in history. For three years, there was a cortisone famine while companies worked out a way to mass-produce it.[44] Even so, there was no detailed understanding of how it worked as a medicine. This was an era when randomised clinical trials had only just begun to be used,[45] and very little was known about the components of the immune system, so demand for the hormone and how to use it – including the dose and type of patient to treat – all came from ad hoc observations, rumours and anecdotes. It's probably lucky that it happened that way. If someone suggested today that a compound which impacts the activity of one in five of all human genes might make a useful drug, nobody would take them seriously. It would sound far too messy and complex to be likely to work.[46]

Hench knew that cortisone wasn't a miracle cure for rheumatoid arthritis, even though some of the press reported that it was. It only gave relief of symptoms for a relatively brief time. 'Cortisone is the fireman who puts out the fire, it is not the carpenter who rebuilds the damaged house,' he said.[47] More importantly, it soon became clear that there were significant side effects when cortisone was taken repeatedly at a dose high enough to help rheumatoid arthritis patients.[48] These included muscle weakness, fatigue and weight gain. But it was then, just as the side effects of using cortisone for rheumatoid arthritis patients became clear, that its true lasting medical importance emerged. It was found that cortisone could treat asthma (as well as some other diseases) at far lower doses than were required for the treatment of rheumatoid arthritis. Since then, cortisone and its derivatives – often just called steroids, the name for this class of compounds with similar chemical structure – have, year after year, been among the world's most widely prescribed medicines.

Cortisol itself is also used as a medicine – in which case it is often referred to as hydrocortisone – for example, in a cream that can be applied to skin to reduce swelling or treat minor irritations. A synthetic chemical very similar to cortisol – dexamethasone – is about forty times more powerful in suppression of immune responses and is used in an enormous number of ways, to treat rheumatic inflammation, skin diseases, severe allergies and more. Other medicines similar to cortisol are used in preventer inhalers for asthma.

It's common for science books which feature medical advances to include anecdotes of patients' stories as an emotive hook to the narrative. Encouraged by my publisher to do this, I asked my son, aged twelve at the time, what he thought of his asthma inhaler. He looked at me as if I had just asked 'Shall we to go to Mars today?' and walked out of the room. He makes a point. Many people with mild asthma no longer need think of themselves as a patient. Inhalers are a part of everyday life: an outcome from one of science's greatest ever detective stories.

Surprisingly perhaps, Hench's and Kendall's scientific careers didn't end in the glory one might expect. Although Hench wasn't formally diagnosed, many, including his son John, thought that he became depressed, or at least that his demeanour changed, after winning the Nobel Prize. When scientists and clinicians criticised cortisone on account of its side effects, Hench took it personally. His son recalls that 'along with other people … who don't draw boundaries between their work and the rest of their lives, Dad found it very difficult to take criticism of his work and accomplishments as anything but disloyalty'.[49]

Hench had planned to write a book about the history of yellow fever. The topic is not as arcane as it might sound: US army doctors proved a Cuban scientist's idea that the disease was carried by mosquitoes, which led to new paradigms in humanitarian medicine and ethics. To dig up the story, Hench applied the same depth of rigour that made him a great biologist. He spent twenty years collecting thousands of documents, photographs and artefacts, and interviewed many of the physicians and scientists involved. The items he collected filled 153 boxes.[50] But he died, aged sixty-nine, the book unwritten.

For Kendall too, his past success proved to be no guarantee of future triumph. It is perhaps telling that his memoir, published in 1971, ends with his winning of the Nobel Prize in 1950.[51] Soon after winning the prize, he was forced to leave the Mayo Clinic because of their strict policy that staff should retire at age sixty-five. He moved to Princeton, New Jersey, where he focused on searching for another adrenal hormone, one postulated to be like vitamin C. He spent twenty years searching for it, but it didn't exist. Success – even at the highest level, discovering one of the world's most important medicines and then winning a Nobel Prize – is essentially fleeting.

*

As well as being one of the world's most important medicines, the discovery of cortisol opened up the molecular basis for how our mind and body are connected. 350 years after Descartes theorised the separation of mind and body, cortisol brought them together, showing how a mental experience – stress – results in physiological effects. Understanding the full implications of this connection between our mental state and our immune system is an especially fascinating but controversial subject of ongoing enquiry.

Our modern understanding of stress began in 1936 when Hans Selye, born in 1907 in Vienna and then working at McGill University in Montreal, discovered that rats exposed to different types of harmful situations – surgery, drugs or cold temperature – showed a similar physiological response, independent of the precise nature of the situation.[52] At first his work received little attention but he soon gained fame and was nominated for a Nobel Prize several times.[53] By the time he died in 1982, aged seventy-five, he had published 1,600 articles and thirty-three books about stress.[54] Selye took stress to be 'the nonspecific response of the body to any demand'.[55] Or, as he wrote in one of his bestselling books: 'The soldier who sustains wounds in battle, the mother who worries about her soldier son, the gambler who watches the races – whether he wins or loses – the horse and the jockey he bet on: they are all under stress.'[56] When Selye was asked if he thought modern life had become too stressful, he replied: 'People often ask me that question, sometimes comparing our lives with that of the caveman ... They forget that the caveman worried about being eaten by a bear while he was asleep, or about dying of hunger, things that few people worry about much today ... It's not that people suffer more stress today. It's just that they think they do.'[57] And Selye often emphasised that stress is not all bad; that it is also, he said, the spice of life.[58]

As we have seen, stress – whether it is taking exams, relationship problems or strenuous exercise – causes the adrenal glands, situated on top of our kidneys, to pump out hormones including cortisol.[59] Cortisol's function is to prepare the body for a change in activity, and levels of cortisol in a person's blood don't only change with stress; they vary according to the time of day as well. Cortisol levels are highest in the morning, peaking around 7 to 8 a.m., and lowest at night. It's thought that the morning increase helps the body prepare for the change in activity of waking up.[60] Still, cortisol levels change much more dramatically with stress, and in so doing they dampen our immune system. Cortisol does this by reducing the efficiency with which immune cells engulf germs, produce cytokines or kill diseased cells. This is fine for a brief time, but if stress persists, our immune system may stay weakened.

There is evidence that people who are stressed for prolonged periods of time suffer worse from viral infections, take longer to heal wounds, and respond less well to vaccination.[61] All kinds of stresses have been linked with diminished immune responses, from burnout at work to unemployment. Even natural disasters like a hurricane can alter the state of people's immune system.[62] Well over a hundred clinical studies have reported that stress can contribute to poor health, which leads many to suppose that a super-charged lifestyle perhaps increases our risk of all kinds of illnesses, from autoimmune disease to cancer.[63] The topic remains controversial, however, because so many factors affect our ability to fight disease that it is difficult to assess the effect of any one.

To explore the relationship between stress and health, without the added complication of stressed individuals being more likely to exercise less, sleep poorly, drink alcohol or smoke, for example, some researchers turn to studying mice where the variables are more easily controlled.[64] Mice can be stressed by placing them in a tunnel in which they can freely run up and down, but are not able to turn around. This type of restraint, applied overnight when mice are most active, causes dramatic changes to the immune system. When given a dose of flu, there's a delay in the immune response of stressed mice. Lower numbers of immune cells move into the infected lungs, and cytokine levels are lower.[65] If the stressed mice have been given a drug which blocks the effect of cortisol beforehand, their immune system responds normally. This is strong evidence that stress and immunity are directly linked through cortisol levels. Similarly, rats stressed by predator odour or swimming in cold water are weakened in their ability to control a candida fungal infection.[66]

In humans, elderly people stressed by caring for a spouse with dementia have a reduced response to a shot of flu vaccine.[67] There is also evidence that stress can affect our response to HIV. Our immune system can keep the virus in check before eventually AIDS develops, but the length of time it can do so varies between people. Over a five-and-a-half-year period of study, it was found that the probability of men infected with HIV developing AIDS increased two to three times if they had higher than average stress, or less social support.[68] A separate study of homosexual men came to the conclusion that AIDS advanced more rapidly in men who conceal their sexuality, although the reasons for this were not established.[69] Many other studies have found that stressed individuals are more prone to reactivation of herpes.[70] Overall, the bad effect of stress on health is probably the best-established link between lifestyle and the immune system.

As well as stress, other states of mind likely also affect our immune system, although the evidence is less robust.[71] Rugby players, for example, increase the level of cytokines in their blood when feeling angry or aggressive just before a match.[72] This fits with the idea that, since aggression often precedes violence, a heightened immune system would be beneficial to deal with germs that enter wounds. Laughter may also help boost the immune system. People with diabetes who watched comedy films with hospital staff experienced increased activity of their immune system genes.[73] This might be due to laughter itself, or perhaps the social camaraderie of laughter.[74] The effects of laughter on the body are, in general, very little understood.

While many emotions *may* impact the immune system, only the influence of stress is widely accepted, which then raises the question of whether or not practices that might reduce stress – from adult colouring books to psychoanalysis – could directly boost our immune defences. There are any number of ways to relax, but two examples that have been studied for their effect on the immune system are t'ai chi and mindfulness.

Practitioners of t'ai chi, developed as a martial art in China, or related exercises such as qigong, perform a slow meditative choreography of movements. There is good evidence that t'ai chi can help improve pain and physical mobility for elderly arthritis patients.[75] Whether or not t'ai chi impacts the immune system, however, is controversial. In one study, a t'ai chi class taken for an hour three times a week led to elderly adults responding better to a flu virus vaccine.[76] This is an interesting result but this type of research is often less definitive than it might seem at first. One problem is that studies such as this often involve only small numbers of people. In this particular study, only fifty people were tested – twenty-seven people who took t'ai chi classes were compared to twenty-three who didn't. Other studies testing for a link between practising t'ai chi and health test similarly small numbers of people.[77] This is something like the number of people who would be enrolled in the first phase of a clinical trial for a new pharmaceutical drug, merely to test the safety of the drug, not whether it works. For a drug to be approved as a new medicine, it is usually tested in thousands of people, and compared to other interventions.

A second problem is bias. In around half the trials testing the effects of t'ai chi on immune defence, it's not clear whether those who took the t'ai chi classes, as opposed to those who did not, were selected randomly.[78] If those who took the classes had already been doing so before the study began, and were merely picked out as a group of people practising t'ai chi, then there is no way of knowing if the effects observed are owing to the classes themselves or to some other shared characteristic that also happens to result in people taking up t'ai chi. More subtly, the control group – those subjects who don't take the class – should be given another activity to perform to replicate the possible benefits of a t'ai chi class that are not actually to do with t'ai chi itself, such as the contact time with a social group.[79]

A third problem – probably the most difficult to address – is how one measures the results. In the study mentioned, testing the effects of t'ai chi on the elderly, what was actually measured was the amount of antibody present in a person's blood after they had been given a flu vaccine. While this might point towards t'ai chi having had an effect on the immune system, we don't know, for example, whether a particular level of antibody increase is sufficient to appreciably impact a person's well-being when infected with flu. The reason that this is such a difficult problem to address is that it's not so easy to design an ethically sound trial to test for our reaction to an actual illness, because the illness would then need to be given deliberately.

Overall, a review of sixteen clinical trials concluded that: 'Because of methodological flaws in existing studies, further vigorously designed large-scale placebo-controlled, randomized trials are needed.'[80] Another analysis of thirty-four trials looking at the effects of t'ai chi, qigong, meditation and yoga came to a similar conclusion: that these practices can have a positive effect on some markers of the immune system, but there is not enough information to determine whether or not immunity is improved against a real infection.[81] Both the National Institutes of Health in the USA and the National Health Service in the UK advise that t'ai chi *may* have various health benefits.[82]

That it *may* help is all we really know about a lot of things. When our children wanted my wife and me to buy them a home trampoline, they found plenty of evidence to demonstrate that bouncing on a trampoline has great health benefits – and claimed that this had been proven by NASA. Impressed by NASA's involvement, I looked into it. As it turned out, the study in question was not as rigorous a programme of effort as that undertaken to land a man on the moon, involving as it did only eight students as subjects.[83] Not only were there so few subjects, none of whom were women, but all eight students were given the same Nike shoes to wear. Would the results have been different if they had worn different shoes, or no shoes? No single study is definitive; it's essential that studies are repeated by other scientists and the results successfully replicated if we are to trust them in order to rule out the possibility that the particulars of one experiment affected the results. We pointed out to our children that there are also safety risks with home trampolines to be balanced against any benefits.[84]

In the end, it's your call (or your parents') as to whether or not you buy a trampoline. And in the case of some things that are supposed to offer health benefits – such as a trampoline – it seems right that it's left to you to decide. But we don't want to have to decide for ourselves about pharmaceutical drugs; we want rigorous clinical trials to inform us if and when they work. Practices like t'ai chi fall somewhere in between.

What sets t'ai chi apart from bouncing on a trampoline is that it provides not just a method of exercise but a narrative for health. There is a story to the movements of t'ai chi; practitioners talk about moving energy around the body to balance one's chi. The power of story is often part of a cure. It's why naming a condition is important, and why a physician's bedside manner, their description of an illness and how they intend to deal with it, can have such a major impact on how patients respond. This power of t'ai chi – the power of its narrative – is hard to quantify.

To take another example, there has recently been great interest in using mindfulness, a non-religious form of meditation, to improve health. Developed in 1979 by Jon Kabat-Zinn – the son of an immunologist[85] – at the University of Massachusetts Medical School, mindfulness uses attention-focusing techniques to instil a moment-to-moment awareness. As Ruby Wax, comedian, writer and mindfulness practitioner, puts it: 'Mindfulness is a way of exercising your ability to pay attention: when you can bring focus to something, the critical thoughts quieten down.'[86]

A review of forty-seven trials testing a total of 3,515 participants concluded that mindfulness can indeed ward off the negative effects of stress, anxiety, depression and pain. The effect is small but similar to what is often achieved with an antidepressant drug.[87] In one clinical trial directly comparing mindfulness with antidepressants, both improved the well-being of patients with recurrent depression to a similar extent.[88]

As well as helping people cope with depression or anxiety, mindfulness is practised more widely as a way of dealing with everyday stress. To enthusiasts, mindfulness is the ideal antidote to the pre-eminent problem of our age: distraction.[89] It might be assumed that, like t'ai chi, reducing stress by practising mindfulness could lower a person's cortisol levels and, in turn, boost the immune system. And in 2016, an analysis of twenty trials with a total of 1,602 participants tested this exact idea.[90]

It was found that mindfulness could indeed lower some markers of inflammation and increase numbers of particular T cells in HIV-diagnosed individuals, but other measures – levels of cytokines or antibodies in blood, for example – were affected in some trials and unaffected in others. The authors concluded: 'we caution against exaggerating the positive effects of mindfulness meditation on immune system dynamics until these effects are further replicated and additional studies are performed'.[91] In fact, it's not actually clear whether mindfulness impacts cortisol levels at all; different trials come to different conclusions.[92] Unsatisfactory as it is, all we know is that mindfulness *may* help.

One of the reasons we don't know for sure if t'ai chi or mindfulness can boost the immune system is that the cost of finding out is prohibitively high. In general, a clinical trial big enough that it would lead to FDA approval if the results were positive costs around $40 million. Eyeing up the possible profits, pharmaceutical companies are willing to pay such sums to test their novel compounds. But who would, should, or could pay to test an unpatented practice like t'ai chi?

While the medical importance of cortisol, and its derivatives, is clear, there remains much more to be understood in how our body, brain and behaviour each affect one another. Evidently, our immune system is a realm of interaction not just between the body and other organisms but also between the body and the mind, and between our physical and mental well-being. And as we shall see next, it also connects us to the solar system itself.

6 Time and Space

The light of day and dark of night bind our lives to time. Because we are moved by the continuous rotation of the earth to face the blazing of the sun and the blackness of space in turn, almost everything about us follows a twenty-four-hour cycle, tuning our lives to the predictable oscillation of our environment. In fact, all life on earth – animals, plants, bacteria and fungi – follows this same rhythm; a rhythm that the rotation of the earth established in living systems probably around 2.5 billion years ago.[1]

The activities of our genes, proteins, cells and tissues wax and wane not simply up or down according to whether it is day or night, though, but each in its own particular cycle, bringing the human body through all manner of peaks and troughs, wave upon wave. Our deepest sleep is around 2 a.m.; the body is coldest at 4.30 a.m.; at 8.30 a.m. secretion of testosterone is at its highest; our reaction time is fastest at 3.30 p.m.; and at 6.30 p.m. our blood pressure peaks.[2] Apparently the best time for sex is 10 p.m.[3]

The body's daily rhythm affects our well-being in all kinds of ways. Accidents at work happen more often at night.[4] Car crashes peak around 3 a.m.[5] Truly catastrophic events, like the Chernobyl nuclear disaster and the *Exxon Valdez* ship oil spill, also tend to happen at night. Perhaps this is because our alertness dips at night.[6] Outcomes of surgery can vary according to the time of day too. Patients are more likely to experience problems if their treatment begins (i.e. anaesthetics given) in the afternoon – although such problems are limited to slight increases in vomiting or pain after an operation rather than anything more serious.[7] It is, however, hard to be certain of the underlying causes behind these kinds of observations. The success or otherwise of a surgical intervention, for example, could be affected by the surgeon's stress or tiredness, the scheduling of more difficult operations at a particular time of day, a change in the patient's capacity for healing, or the inherent peak and trough in the surgeon's alertness caused by the circadian rhythm of the human body.

To test if and how the time of day directly affects the immune system, one needs to eliminate these many variables, and to do so many scientists turn to studying animals. There is considerable evidence that the immune response of mice to an infection is dependent on the time of day at which the infection is contracted. Mice are nocturnal. A stronger immune response is triggered in mice given a dose of salmonella bacteria at 10 a.m., early in their rest time, and a lesser immune response is triggered if the animals are infected at 10 p.m., just as they are becoming more active.[8] In a separate experiment, mice infected with bacteria which cause pneumonia again reacted most strongly when infected in the morning.[9] The mouse immune system, or at least its ability to react against these particular types of bacteria, is more potent during the day, when they are usually at rest. And roughly speaking, the same is true for us in that our immune system is stronger during our natural rest time, at night.

One reason for this is that, as we saw in the previous chapter, the immune-suppressive hormone cortisol is kept low during the night. Another is that a greater number of many types of immune cell flow through the blood at night-time. For a few types of immune cell, however, the complete opposite is true – some types of T cell, for example, are more abundant in blood during the day[10] – so the idea that the immune system is simply better or worse by day and night is too crude. Although it provides less of a sound bite, it would be more accurate to say that our immune system is in a *different state* depending whether it is day or night.

It is possible that our immune system has evolved this way to deal with germs that attack us at different hours. Many disease-causing mosquitoes are more active at night, for example, so there could be a benefit to our night-time immune system being tuned to handle malaria parasites and other germs carried by mosquitoes. A complication with this idea is that some mosquitoes are active during the day, those that transmit dengue and Zika viruses for example, and some of the malaria-transmitting mosquitoes in Asia also bite during the day or at dusk. The idea that our immune system changes its state to fight different types of germs by day and at night is difficult to test directly.[11] It's even possible that, rather than helping us, our body clock could work against us here. Parasites that rely on insects that bite for their transport from one body to another might be able to use a night-time change in the body as a cue to congregate in the skin, ready for the insects' arrival.[12] A bird parasite found in turkeys is known to change its location in the animal ready for being picked up by insects.[13]

Another possible reason for our immune system behaving differently during the day and night is simply that, as one scientist puts it, 'it has no choice'.[14] 'The function of sleep,' researcher Till Roenneberg puts it, 'is to make us fit for being awake.'[15] From this perspective, our immune system may not have evolved to respond differently by day and night for any particular benefit but as a side effect of the body's twenty-four-hour cycle which has evolved to optimise the body's use of energy. It is probably the current consensus that 10–15% of all our genes vary their activity by day and night, primarily to regulate the body's metabolism, and as a consequence of this, all of the body's processes are affected, including the immune system.

Whether or not these changes have evolved to benefit the immune system per se, they have many consequences. Symptoms of asthma, caused by an unwanted inflammation of a person's lung airways, are more common at night, for example, and sudden deaths due to asthma often occur around 4 a.m.[16] Gout, an inflammation of the joints caused by an unwanted immune response triggered by a build-up of uric acid crystals in joints or on tendons, also worsens during the night.[17] Patients with rheumatoid arthritis, on the other hand, tend to suffer worse symptoms – stiffer joints, for example – in the morning. This correlates with immune-stimulating cytokines building up through the night while the suppressive hormone cortisol is low.[18] Illnesses less directly associated with our immune system are also affected: for reasons that aren't clear, migraine headaches often peak at a particular time of day, most commonly in the morning, and toothache peaks around 9 p.m.[19] Sudden cardiac death is more likely between 9 and 11 a.m., and seizures caused by temporal lobe epilepsy occur more frequently between 3 and 7 p.m.[20] In short, many disease symptoms vary according to the time of day or night but there's no simple rule how; each is affected differently.

Our body's twenty-four-hour cycle is such an integral and influential aspect of our physiology that disruptions to it can be harmful. Jet lag, as everyone knows, is more than tiredness. It is caused by the body having to adjust to a new schedule of light and dark, activity and rest. Repeated jet lag has been simulated in mice – by bringing their artificially controlled 'daylight' hours forward by eight hours every other day for ten days – and their health suffered as a consequence; tumours grew faster in jet-lagged mice and their survival from cancer decreased.[21] In us, long-term night-shift work has been linked to an increased risk of breast cancer.[22] But this relationship has not led to specific regulations or guidelines being drawn up because the issue is still open to debate: an increased risk of breast cancer was only apparent for people working night shifts for thirty years or more, and other factors, such as a lack of exercise, could feasibly account for the increase.[23]

To simulate the impact of night-shift work on people in a more controlled situation, volunteers lived for six days in a laboratory in which their normal sleep habits were delayed by ten hours.[24] This changed some aspects of the volunteers' immune system but not others. The peak time at which immune cells secrete cytokines changed, for example, but particular types of immune cells were still more abundant in the blood at night.

One reason for such complex consequences is that there isn't just one timing system in the human body. There are several clocks running simultaneously. For the most part they are synchronised, but each has its independent mechanisms and can fall into a rhythm of its own.

The body's master clock, which acts as the conductor of the orchestra, is made up of around 20,000 nerve cells located in the hypothalamus at the base of the brain. This in turn takes its cue directly from our eyes. Millions of light-sensitive cells at the back of the eye, called rods and cones, capture a mosaic view of the outside world, providing our vision, but the master body clock in the hypothalamus is tuned to something else. In 1991, an experiment hinted that eyes can relay information to the brain about the world beyond what it looks like. It was discovered that mice whose rods and cones didn't work properly – so that they were blind – were still able to regulate their body clock according to cycles of light and dark.[25] Russell Foster, who made this discovery, then at Imperial College London, proposed that this meant there must be another type of cell in the eye, different to rods and cones, whose purpose was not to help form an image of the world but whose photoreceptors specialised in detecting simply *how much* light there is – brightness – for the purpose of controlling the body clock.

The scientific establishment reacted to his idea with something like: 'Look, we've been studying the eye for 150 years, are you seriously telling us that we've missed something this big?'[26] Foster's grant applications to test the idea were rejected. Once, someone in the audience at one of his talks shouted 'Bullshit' before walking out. But Foster found solace in the example of Thomas Henry Huxley – who took on the establishment throughout the mid-nineteenth century in order to champion the ideas of his friend Charles Darwin. Eventually, evidence accumulated in Foster's favour, like it did for Huxley before him, and everyone came to accept his once-heretical idea that a small number of cells at the back of the eye aren't there to form a picture, but instead respond to the brightness of our environment – information which the brain uses as a twenty-four-hour metronome.[27]

(I first met Foster soon after his ideas had been proved right, in 1999, when I faced the task of setting up my own new lab two floors up from his in Imperial College London. His advice was that I apply for as much research funding as I possibly could, and then some, with urgency. You never quite know what will be approved, he told me. Foster often said that having his idea proved right in the end 'reinforced the sense that you must keep kicking on the bloody door till the buggers let you in'.)[28]

As Foster discovered, the hypothalamus clock takes cues from special cells in the eye, but this doesn't drive the body's rhythm all on its own. While the hypothalamus clock acts as conductor, harmonising the ensemble, the players – all the rest of our body's cells and tissues – are quite capable of keeping their own time, as their own genes and proteins wax and wane. Even red blood cells, which have no nucleus or genes of their own, oscillate their activity for many days without any external cues.[29] The problem for night-shift workers is that as they alter the timing of activity, digestion and sleep, affecting the clocks that are running in various tissues and organs, the master clock in the brain continues to follow the light and dark of day and night: the conductor falls out of sync with the ensemble. We can recover from, say, jet lag when all of our body clocks shift to adopt a new twenty-four-hour cycle, but there's no simple way for the body to get used to night-shift work because clocks running in some tissues and organs will always be out of sync with the hypothalamus clock.

The consequences of disrupting our body clock are especially evident when magnified by the extreme conditions of space. As the International Space Station whizzes around the earth at around 17,000 miles per hour, astronauts are in sunshine for forty-five minutes, then darkness for forty-five minutes; sixteen days whizzing by for every one of ours on earth. A survey of sixty-four astronauts on space-shuttle missions, and twenty-one astronauts on the International Space Station, showed that most take drugs to help them sleep.[30] Blood taken from space-station astronauts several times over a six-month period showed that, by all kinds of measures, their immune systems were in disarray.[31] Many types of immune cells were redistributed in the body, activation thresholds had shifted and T cells had become less responsive.[32]

As far as we know, nobody has developed cancer or an auto-immune disease in space.[33] NASA has its own rule that astronauts should not have their lifetime cancer risk raised by more than 3% by their work.[34] Contrary to popular belief, though, astro-nauts do experience medical problems in space.[35] These tend not to be from recent infections; since the earliest missions, precautions have been taken to prevent such occurrences. (Three days before Apollo 13 launched on 11 April 1970, command module pilot Ken Mattingly was replaced by back-up pilot John Swigert, because it was discovered that he had been exposed to the same measles virus that had infected another Apollo astro-naut, Charlie Duke. Mattingly was fortunate: he avoided the mission which famously had to abort its moon-landing after an oxygen tank exploded, but later went to the moon on Apollo 16.) Instead, it is relatively common for a virus that is already present in the astronaut's body but which is dormant to be reactivated – in a similar way that the chicken-pox virus can reactivate later in life to cause shingles – probably because the astronaut's immune system is no longer able to keep it under control. Reactivations of all kinds of viruses (cytomegalovirus, Epstein–Barr virus and herpes viruses) have been documented in astronauts on both short- and long-duration missions.[36] As far as we're aware, this has not led to anyone developing clinical problems in space – in other words, while the virus has become active and multiplied, the astronaut displays no symptoms of illness – but medical privacy or other rules about space flight might prevent disclosure of such a problem even if it did occur.

As well as reactivation of latent viruses, several crew members of the space station have developed a skin rash. In one case where blood samples were analysed, the presence of a rash correlated with changes in their immune system, including reduced functionality of their T cells and altered levels of cytokines in their blood.[37] For this astronaut, the rash coincided with itchy watery eyes and sneezing, indicating an allergic reaction. This almost certainly developed from a disruption to the immune system caused by space flight. The astronaut had never had any of these problems on earth before and symptoms cleared up within days of the astronaut returning to earth. Symptoms peaked at the same time as stressful events during the mission, including just after a spacewalk, which fits with the idea that stress tends to make allergic reactions worse.[38]

Allergies in space are not rare. Antihistamines, used to counteract the effects of allergies, are the second most-taken medicine in space after sleeping pills.[39] In at least one case, supplies of antihistamines ran out and more had to be sent up in the next scheduled space-shuttle docking. So for long-term space missions, the prospect of allergies, reactivation of latent viruses and possibly the development of autoimmune disease or cancer, are serious concerns. Brian Crucian, in his 'dream job' as NASA's lead scientist for all things immunology, thinks this could be a problem for, say, a trip to Mars. But, he says, it's hard to say if it's any more of a problem than the many other effects space flight has on the body: in addition to changes to the immune system, it undergoes bone loss, muscle loss, cardiovascular problems, impaired vision and psychological stress.[40] In short, we are not built for space. The human body has evolved to fit our environment. It is tuned to the level of gravity felt at the earth's surface, the twenty-four-hour cycle of day and night, the way we interact socially, and so on. If there is ever a realistic plan for humans to settle elsewhere in the solar system, our immune system and many other bodily systems will need to be tricked into thinking we haven't left home.

Nonetheless, when Crucian talks about his research, there's excitement in his voice because even if the challenges of long-term space flight are great, we are learning much about human health in the process.[41] Studying health problems caused by stress, isolation, confinement, changes in nutrition, exercise, sleep or unusual cycles of light and dark – either in space itself or at places such as the Concordia research station in Antarctica, which is in complete darkness for four months of the year and 600 kilometres from the nearest human contact – may well lead to medical benefits for us all. New cures often come from research institutes, pharmaceutical companies and university medical schools, but they can also come from left field – from projects such as NASA's to explore space.

Our understanding of the body's daily rhythm already opens up an opportunity – to do with the timing of medication. As symptoms of diseases and the activity of our immune system change during the day and night, it follows that medicines might best be given at specific times of the day. For asthma patients, an inhaled steroid given once a day between 3 and 5.30 p.m. was shown to be superior to giving the same medicine at 8 a.m. In fact, it was comparable to taking the medicine four times a day.[42] Statins, widely prescribed to prevent heart disease by lowering cholesterol, are often taken at night when most cholesterol is produced (though how important this is actually depends on the particular statin prescribed).[43] Timed delivery of medicines might be more widely important than current practice suggests, as fifty-six of the top hundred bestselling drugs in the USA, including all of the top seven, target the product of genes that change their activity with the time of day.[44] Around half of these bestselling drugs stay active in the body for only a short time after being taken, so matching the time they are taken with when they will be most effective may well improve their performance.

One problem with giving people a narrow time window in which to take their medicine is that around half of people with long-term illnesses already don't take their drugs as prescribed.[45] Complex timed regimens are unlikely to be adhered to. One solution is to automate the delivery of the medicine. In the near future, soft stretchable gel-like plasters with microchips and drug delivery channels embedded in them might be able to administer medicines in a programmable way, perhaps even in response to readings taken from the body, such as skin temperature.[46]

For now, vaccination is one area where timed delivery is easily implemented.[47] One study found that stronger immune responses were generated for vaccinations against hepatitis A or flu when the vaccines were given in the morning compared to the afternoon.[48] In this small trial, however, the timing of the vaccine's administration was not randomised, so there could be other factors at work: perhaps there are particular characteristics of people who prefer to be vaccinated in the morning that also affect their immune response to vaccination, for example. Bizarrely, morning vaccination was found to be beneficial only for men; women responded to the vaccine similarly whether given in the morning or the afternoon. It is possible that the rhythms of the human body have different consequences for the immune system of men than they do in women, but this has never been studied directly.[49] Overall, the usefulness of vaccinating at specific times of day is not yet widely accepted. Another small study, for example, showed that a hepatitis B vaccine was equally efficient whether given in the morning or the afternoon.[50] Generally speaking, small studies can make any emerging idea look either fanciful or promising: larger-scale studies are needed to really test whether or not these simple changes in medical practice could have significant benefits.

While some scientists test whether or not existing vaccines work better or worse at different times of the day, others are striving to create vaccines that, by design, take advantage of the body's daily cycles. We have already seen how vaccines can be improved by using adjuvants which target specific immune cell receptors, such as toll-like receptors, to trigger a reaction. Such vaccines might be made more effective still if given at the time of day when a toll-like receptor is especially responsive. This has already proved to work in mice.[51] Mice given a vaccine containing an adjuvant that works through a particular toll-like receptor reacted best when the vaccine was given in the middle of the night, the optimal time of day for that toll-like receptor to respond. Weeks later, mice vaccinated at night still showed improved immunity. It seems likely that something like this will be true for humans. Even a small improvement in the effectiveness of a medicine that can be achieved like this, with relatively little cost, might lead to many hundreds or even thousands of people living longer or more healthily. In the end time kills, but in the interim, it could be part of the cure.

*

Our relationship with time is changing thanks to one of the greatest triumphs of humankind in the last century, perhaps in all history: our increased lifespans. In East Asia, where this change has been most dramatic recently, life expectancy for those born in 1950 was forty-five but today is over seventy-four.[52] Some of this increase in life expectancy can be accounted for by decreased child mortality, but we are also, on average, living far longer. In the UK and USA, the number of people over ninety years old has trebled over the last thirty years.[53] This presents us with a new problem: to improve our lives in old age so that we don't just live longer but stay healthy and active for longer too.

In the US, those over sixty-five make up 12% of the population but account for 34% of prescription drugs and 50% of hospital stays.[54] Partly this is because as we age our bodies become weaker at fighting infections; 80–90% of those who die from the flu virus, for example, are aged sixty-five and over.[55] The situation is not helped by the fact that the elderly also respond less well to vaccines.[56]

It's not that our immune system simply becomes unresponsive as we age, because the elderly are also far more likely to suffer from autoimmune diseases, caused by unwanted immune responses. Rather, it seems that our immune system somehow goes awry as we age. Given that everyone on earth now aged sixty years can expect to live on average another twenty years, understanding what happens to our immune system as we age is a hugely important frontier of science. So, what is ageing exactly?

No matter how much we rage against the dying of the light, ageing is inevitable – right down to the level of our constituent cells. In a lab culture dish, adult skin cells divide roughly fifty times before stopping, and skin cells from a newborn baby can divide eighty or ninety times; cells from someone elderly, on the other hand, divide only around twenty times. Ageing is even evident in our genes. Our genetic material is modified over time – chemicals can become attached to it, and the way the strands of DNA are folded up can also be altered – changing which genes are easily switched on or off. These processes underlie what is called epigenetics, the modification of genetically encoded traits by one's environment. Another kind of change takes places at the end of our chromosomes, where repeating segments of DNA called telomeres are found. Telomeres act like the plastic tip of a shoelace, preventing the twisted coils of genetic material from fraying at the ends or knotting together. But telomeres shorten each time a cell divides.[57] We don't know if short telomeres are merely a mark of ageing, like grey hair, or are part of the process by which cells age. It is possible that telomeres work as a tally of how many times the cell has divided so that the cell knows when to stop.

The situation is complicated by the fact that some cells can also increase the length of their telomeres, using an enzyme called telomerase. In fact, immune cells use this enzyme to stop their telomeres shortening when they multiply, as do cancer cells – likely a contributing factor in their apparent immortality[58] – and drugs that stop telomerase from working show promise against cancer (although cancer cells can evolve resistance). There is also evidence that stress can influence telomerase activity,[59] though this is perhaps not so surprising given the enormous number of effects stress has on the human body. Supporting the potential for stress-reducing interventions to improve health, at least one study has found that practising mindfulness correlated with improved telomere maintenance in breast-cancer patients.[60]

Given that ageing has such a profound effect on our cells and genes – the effects mentioned here being just some examples – a much bigger question emerges: why does this happen, why do we age at all? It was once thought that ageing, leading to death, evolved as a mechanism for ensuring the continuing evolution of species. For any species to evolve – for features of a species to change over time – there needs to be a turnover of individuals. However, one problem with this idea is that most life on earth doesn't ever reach old age. Most animals are killed by predators, disease, the climate or starvation, so an inbuilt limit on an animal's lifespan isn't likely to have much effect. Another view is that ageing is simply a side effect of the damage that builds up over time in our genetic material caused by, among other things, reactive oxygen molecules produced during metabolism or exposure to UV light. But while it is well established that the extent to which our genes are damaged does increase as we age, it is not proven that this drives ageing directly. The fact that our genes become increasingly damaged over time leads, however, to another possibility – that ageing might have evolved as a form of defence against cancer. Since cells accumulate genetic damage over time, they may have evolved a process to not persist in the body for too long, in case this damage eventually causes a cell to turn cancerous.

Cancer occurs, after all, when a cell multiplies excessively, while ageing, on the other hand, is the process by which a cell does the opposite, either entering a programme of events called apoptosis that leads to the cell's death[61] or entering a state called senescence, in which the cell stays alive but no longer multiplies. Senescent cells accumulate in the body over a lifetime – especially in the skin, liver, lung and spleen – and have both beneficial and detrimental effects.[62] They are beneficial because they secrete factors which help repair damaged tissue, but over a long period of time, as senescent cells increase in number, they can disrupt the normal structure of organs and tissues. These cells could be an underlying cause of many of the problems we associate with ageing. Mice in which senescent cells were cleared were profoundly delayed in showing signs of ageing.[63] Even mice already showing signs of ageing could improve their muscle structure and fitness when senescent cells were cleared.

A final possibility worth mentioning here is that the genes which make us age might be passed on from one generation to the next simply because they have some positive benefit to us when we are young and their negative side effects are manifest only after we have reproduced, so aren't strongly selected against.[64] Overall, we can describe a lot of what happens during ageing, at the level of what physically happens to our genes, cells and organs, but the fundamental question of *why* we age is still open. In all likelihood there is more than one correct answer. (Don't listen to anyone who says the big questions have already been answered.)

Returning to the immune system specifically, part of the problem we face in old age is that our bodies produce fewer immune cells. Different studies have come to slightly different conclusions as to which types of immune cell continue to be produced at a steady rate and which do not, but there is a decrease overall. One of the probable reasons for this is that bone-marrow stem cells, which produce immune cells, lose their regenerative potential over time likely because of the accumulated damage to their DNA. Evidence for this comes from the fact that bone marrow from an older person is less efficient in establishing a new set of immune cells when used for transplantation to help a cancer patient.[65] This is one reason why charities which recruit bone-marrow donors are especially keen to sign up younger people.[66] In addition, immune cells in elderly people are poorer at detecting signs of disease and respond less efficiently to the protein molecules that direct them to a wound or the site of an infection. Although they are able to move just as fast as cells isolated from younger people, they are therefore less accurate in getting to where they are needed.[67]

This fits with a simple view that the immune system in an elderly person is weaker, but this isn't the whole story. At the same time, signs of an active immune response – cytokines, clotting factors and other inflammatory molecules – are often found at higher levels in the blood of elderly people, even when there are no overt signs of infection.[68] This phenomenon is sometimes called 'inflamm-ageing'. There are many reasons why a low level of background inflammation might persist in elderly people, such as there being an accumulation of damaged or senescent cells, but the effect is that the system is less able to discriminate between germs and the body's own cells and tissues, and is particularly weak at detecting germs that have never been encountered before. In broad-brush terms, it is easier for an immune reaction to be triggered in elderly people but, by the same token, the system is less stringent in responding appropriately.

Some changes to our health might be the inevitable consequence of immune cells ageing, in the same way that all cells age, but this does not obviously account for all of the complicated changes that happen to the system overall. Once again, to understand complex phenomena such as what happens to the immune system as we age, understanding each component of the system helps but doesn't give us the whole picture; we need also to understand how the various components interact. Effects arise not just from the ageing of immune cells but from the ageing of the system as a whole – a consequence of the system already having spent decades battling germs.

As we have seen, each time we fight an infection our body retains some of the immune cells best equipped to tackle that infection, in case we encounter the same germ again. These long-lived cells – our memory immune cells – account for the body's ability to fight off infections more quickly a second time round. This is why vaccines work and why, if you catch a virus, say the flu virus, you would be well equipped to fight the exact same flu virus again (but not necessarily next season's flu, in which some of its genetic make-up will have changed). Crucially, this means that older people have more of their immune cells devoted to battling previously encountered infections, leaving fewer immune cells available for the fight against new infections.

Adding to this problem, the organ where many of our newly made immune cells must develop doesn't work well in elderly people. This is the thymus, which sits in the chest between the lungs, where T cells develop before patrolling the body to look for signs of disease. Recall that T cells have receptors with randomly shaped ends giving them an ability to lock onto, and react against, all kinds of other molecules. Any T cells which have, by chance, a receptor that could trigger a reaction against the body's own healthy cells are killed off in the thymus. T cells allowed to patrol the body are therefore those T cells which do not react to the body's own cells and tissues, and are poised to detect molecules alien to the body, such as components of germs. Unlike most organs, however, the thymus is at its largest in childhood. This is because our immune system develops most dramatically when we are young, each of us being born with only a temporary defence borrowed from our mother which must be replaced by our own immune system. From puberty onwards, the ability of the thymus to scrutinise newly made T cells begins to decline and the thymus itself shrinks in size. It was once thought that the thymus had shrunk so much by the time we reach old age that it no longer allowed for the development of new T cells at all – but we now know that this is not quite true. It does retain some activity.[69] The thymus in elderly people probably works at something like 1–5% of its activity in childhood.[70] It's as if the body has decided that after puberty, it now has most of the T cells it will need for the rest of life.

Once T cells are seldom produced with brand-new receptors, the body's set of T cells is shaped by the particular suite of germs a person has been exposed to over their lifetime, so that the numbers of T cells able to fight those specific germs is increased. Other factors, perhaps including levels of exercise and stress, likely also shape the immune system as we age. Strong evidence that this is the case – that our immune system is not fixed by our genetic inheritance, but adapts as we age – comes from the fact that genetically identical twins vary considerably in their immune system, especially when older. An international team of scientists led by Mark Davis[71] (no relation) at Stanford University analysed the immune system of 105 sets of healthy twins in over 200 ways, including measuring the level of different immune cells in their blood and the ability of immune cells to secrete cytokines before and after participants were given a shot of flu vaccine.[72] Although it has been long known that each person's immune system varies – the number of different types of immune cells in a person's blood varies hugely, for example – Davis and his colleagues set out to establish how much of this variation was inherited and how much wasn't. They found that most aspects of our immune system depend on non-heritable factors far more than on the inheritance of our genes. It has long been clear that some combination of nature and nurture determines our health, but the fact that nurture plays such a large role in the configuration of our body's defence is surprising.

Viruses such as cytomegalovirus – which is a common infection and often doesn't cause any symptoms, although it can cause problems if contracted by a baby in the womb – have an unexpectedly long-lasting impact on the state of our immune system.[73] The presence of cytomegalovirus in young adults, for example, correlates with them have a stronger immune response to a flu vaccine.[74] Davis's analysis of twins also showed that the immune systems of younger identical twins were far more similar than those of older identical twins. The implication is that as we age, the individuality of our immune system increases. We become more, not less, of ourselves.

This complexity – everyone's unique history – is one reason why it is especially hard to design medicines that work with the immune system in the elderly. But that's not to say it's impossible. One way forward is to tailor vaccines specifically for the elderly. Recall how the body's innate immune system involves toll-like receptors locking onto the telltale signs of germs – such as the LPS molecule on the outer coating of bacteria – to help trigger an immune reaction, and how this knowledge led to the development of adjuvants that replicate those telltale molecules. For the purposes of tailoring vaccines to elderly immune systems, one approach is to choose adjuvants that replicate the types of germ molecules that the elderly respond well to. A molecule called flaggelin, for example, isolated from the wiggly protrusions of bacteria, is one of the few germ molecules easily detected by the immune system in people of all ages. One vaccine designed to protect against the flu virus but which also included this bacterial molecule worked far better than a standard vaccine, both in elderly mice[75] and elderly humans.[76]

The advantages of giving vaccines at particular times of the day could also be especially beneficial to the elderly. One trial focused on testing whether or not a flu vaccine would work better for the elderly if given in the morning or the afternoon. The elderly responded better – their levels of blood antibodies were higher a month after the vaccine – when the vaccine was given in the morning, between 9 and 11 a.m.[77] (Although a previous trial, mentioned earlier, suggested there may be differences in how men and women respond to the timing of vaccination, there was no such difference found in this larger study.) However, the advantage of vaccinating in the morning did depend on which version of flu was used. Vaccination against one type of flu worked especially well when given in the morning, rather than the afternoon, and there was a weakly beneficial effect for a second type of flu, but a third showed little, if any, improvement. Perhaps most importantly, it is not known if the differences detected – an increase in blood antibodies – would necessarily translate into an enhanced capacity to fight a real flu infection, something that is hard to test directly because, as we mentioned in the discussion of the effects of t'ai chi on immunity, it's unethical to deliberately expose groups of people to an illness, even for worthy scientific ends.

Janet Lord, who heads the Institute of Inflammation and Ageing at the University of Birmingham, where this trial was conducted, thinks that in the case of the flu vaccine that saw most benefit from timed vaccination – which is in fact a version of flu that is especially hard to vaccinate against – giving the vaccine in the morning would be able to protect over half of elderly people. A trial with many thousands of volunteers is needed to test whether or not morning vaccination is best at protecting people over an actual flu season, but Lord is quietly confident. She thinks it will soon be clear that at least some types of vaccine should be given at specific times of day, especially for the elderly.[78] Another researcher, Akhilesh Reddy at the Crick Institute in London, agrees. He has found that the innate immune response of cells infected with a virus is ten times more potent in the morning compared to the afternoon and thinks that this is likely to relate to why vaccines work better in the morning.[79]

Even if Lord and Reddy are right, many problems with ageing remain unsolved. This chapter cannot end, unfortunately, by celebrating the discovery of a medicine which turns back the clock on an ageing immune system. Instead, it must end with a call to arms to understand the process better. Much of this book celebrates scientific heroes who have roamed freely, funded essentially to follow their nose, and we are rewarded when a novel process comes to light that is outside the framework of current understanding – such as the discovery of the innate immune system, discussed in Chapter One. Funding risky projects, bold ideas, and the whim of individuals, must not stop. But there is also a place for strategic thinking; for some fraction of scientific research to be directed to topics especially important for society. Lord, as an example, took up the challenge of trying to understand the ageing immune system only because the UK's government agency for biological research targeted funds for the topic.[80] She's not the only one.[81]

Ageing of the immune system and ageing in general is, as we have seen, enormously complex. Tackling this frontier now requires the effort of all kinds of scientists, including immunologists, physicians, mathematicians, computer scientists, chemists, physicists, neuroscientists – and those whose career doesn't fit a label. Of course, we have succeeded at tackling complex problems before. When John F. Kennedy gave his famous 'We choose to go to the moon' speech in Houston, Texas, on 12 September 1962, he set down a challenge that many thought impossible to achieve – to put a man on the moon by the end of the decade. NASA had already informed the president that it would take far longer, at least fifteen years, before they could achieve this goal, and a rocket capable of taking a person beyond an orbit of the earth didn't even exist at the time.[82] Kennedy ended his visionary task-setting speech by saying: 'Many years ago the great British explorer George Mallory, who was to die on Mount Everest, was asked why did he want to climb it. He said, "Because it is there." Well, space is there, and we're going to climb it, and the moon and the planets are there, and new hopes for knowledge and peace are there. And, therefore, as we set sail we ask God's blessing on the most hazardous and dangerous and greatest adventure on which man has ever embarked.'

Well, our own inner space is also there, and we must set sail within it. With microscopes not spaceships, we must explore the systems and subsystems of the human body, and we will find ourselves to be more complex than the moon and the planets. This will bring us new hopes for knowledge and for peace. We will understand human nature, our differences and our similarities. We will understand what we want to cure, and we will create new molecules to be those cures. We must take this journey, not because it is there to be taken, but because we must strive to make human existence more palatable and more fulfilling. Especially, I think, now more than ever, as we age.

7 The Guardian Cells

If there's any single take-home message from all of science, it's that nothing is simple. Everything has depth. It sounds simple enough that our immune system defends us by attacking harmful invaders, but it isn't. There are countless complications. The human body, which the immune system should not attack, changes over time; some bacteria are not harmful and do not require a response; dangerous germs try to avoid being detected, and so on. To achieve this simple-sounding mission – to discriminate between what requires a response and what doesn't, and to deliver the right type of response – the human body has invested heavily in a galaxy of cells, proteins and other components, to create a system as elaborate as anything else we know of in the universe. And sometimes it fails.

As we have seen, one way your body tries to ensure that the immune system doesn't attack healthy cells is that as immune cells are made in the body, from stem cells, they are tested as to whether or not they attack the body's own healthy cells or tissues. Any that do so are killed off before they cause any harm. Only those immune cells that do not attack healthy cells are allowed to roam the body and search for signs of disease. But the process isn't perfect, errors happen and healthy cells and tissues can be destroyed without good reason. This is the problem that underlies autoimmune disease.

There are over fifty different types of autoimmune disease – rheumatoid arthritis, diabetes, multiple sclerosis and so on – affecting around 5% of people, two thirds of whom are female.[1] A big problem in treating autoimmune disease is that symptoms often take a long time to become apparent so that by the time somebody visits their physician, immune cells have often been attacking healthy cells for months if not years already. This makes it hard to work out precisely what triggered the immune system to attack healthy cells in the first place. In some cases, immune cells may have reacted to a protein molecule from a real threat, say a virus or bacteria, and then these same immune cells mistakenly react to a normal protein in healthy cells which just happens to be similar to the germ's protein. But this isn't always the case and there are crucial gaps in our understanding of how the immune system works which are important in developing new medicines for autoimmune diseases.

One reason why autoimmunity has been so difficult to understand is that everything about it is so deeply counter-intuitive. For most of history, the very idea that the human body can attack itself has hardly been considered a possibility. A modern view of disease began with Louis Pasteur's discovery of minuscule microbes and Robert Koch's discovery, in 1876, that microbes can cause disease. This replaced ancient views of disease which invoked, for example, an imbalance of the body's 'humours': black bile, yellow bile, phlegm and blood. As well as the uncountable number of medical benefits that came from this shift in thinking, the discovery that germs can cause disease was also the first step towards understanding the immune system as a means of defending us by discrimin-ating between the body's own constituent parts – self – and other things out there – non-self – an idea explicitly elaborated upon in 1949, by Australian scientist Macfarlane Burnet.[2] A few years later, in 1957, a new word was invented – auto-immunity – to describe the idea that disease might also be caused by something entirely different from germs: by the body attacking itself.[3] In 1964, two detailed volumes totalling 980 pages, the output of an international workshop held in New York, led to widespread acceptance that this new concept for disease might underlie many human illnesses. Auto-immunity was one of the most important surprise discoveries of twentieth-century medicine.[4]

A clue to understanding how and why the body sometimes attacks itself is that the same person sometimes shows symptoms of more than one type of autoimmune disease. Type I diabetes is a relatively common autoimmune disease, caused by immune cells attacking insulin-producing cells in the pancreas, which leads to there not being enough insulin to regulate blood sugar levels. Some people with type I diabetes, however, also have problems seemingly unrelated to a deficiency in insulin production, such as not producing enough thyroid hormones, which regulate the body's metabolism, due to immune cells attacking their thyroid glands, or symptoms of coeliac (pronounced see-lee-ac) disease, which include regular stomach pain and diarrhoea, caused by an immune response to gluten. Not everyone with type I diabetes suffers from these other problems of course, but more people than would be expected by chance. A similar thing happens to animals with autoimmune disease. Mice genetically predisposed to diabetes commonly suffer from symptoms of other autoimmune diseases.[5] The implication is that the underlying cause of autoimmune disease is not necessarily something that happens in any one particular organ, but something that happens to the immune system in general, a weakening of its ability to discriminate between healthy cells and harmful germs.

The Japanese scientist Shimon Sakaguchi thought about this carefully. He realised that finding out what can happen to cause the body to attack itself would lead to a deeper understanding of how the immune system works. This drove him to study autoimmunity as a route to understand the system, rather than to seek a cure for autoimmune disease.[6] But he had to find a way to approach the problem, which is formidable even at a glance. How does anyone begin a scientific journey? By following tracks others have left behind, then branching off.

Sakaguchi began by following a path first taken in 1969 by two other Japanese scientists, Yasuaki Nishizuka and Teruyo Sakakura. Working in Nagoya, they had stumbled across a way of causing autoimmunity in mice. They hadn't set out to study the immune system at all – they weren't immunologists, they were endocrinologists, studying hormones and glands, and were looking to test whether hormones affect the development of cancer. To do this, they surgically removed the thymus gland from mice with the idea of then testing what happens if these mice, lacking whatever hormones the thymus produced, then had cancer. What they discovered had little to do with hormones or cancer.

They found that if mice had their thymus removed when they were just three days old, their ovaries were destroyed.[7] At first, the two endocrinologists, focused on studying hormones, took this to mean that the thymus must secrete hormones essential for the growth of the animal's ovaries. But their intuition, squarely derived from their training, was wrong. Later experiments showed that in fact the animal's ovaries had been attacked by the immune system and that other organs in these mice had been attacked as well.[8] We now know that this is because immune cells (T cells specifically) capable of attacking healthy cells and tissues are normally killed off in the thymus. In animals that have had their thymus removed at a young age, self-reactive T cells were not destroyed, which led to auto-immune disease.

At the age of twenty-six, Sakaguchi joined Nishizuka's lab where this discovery was made. For his PhD research, he carried out the exact same experiment, removing a mouse's thymus, but then branched off to make a seminal discovery. 'I only vaguely remember if I got excited or not,' he recalls,[9] because it took three years, between 1979 and 1982, to carry out the experiment. He first had to create many of the reagents he needed – antibodies to mark out specific types of T cells and so on – which took over a year, and then each experiment took several weeks, while he waited to see what would happen to each mouse he treated. Years of effort that boil down to a few lines with a simple dramatic outcome: first, mice had their thymus removed so that, as before, they developed autoimmune disease. These mice were then given an inoculation of immune cells from a healthy mouse (of the same inbred strain) and amazingly, this stopped the autoimmune disease.[10] Mice could be given a dose of immune cells either before or after their thymus was removed and, either way, the autoimmune disease stopped. In other words, Sakaguchi had discovered a cure for autoimmune disease which would otherwise have been inevitable.

This was a profound discovery, especially for a PhD thesis which usually merits a qualification rather than being hailed as a landmark advance. But it wasn't, as Sakaguchi knew, a breakthrough which could immediately inform medical practice, because immune cells can't easily be transfused from one person to the next (genetic differences make this difficult in people, whereas it is easy between inbred mice) and, in his experiment, the mice contracted their autoimmune disease through nonnatural means – a surgical intervention to remove their thymus. Rather than being a medical breakthrough, the importance of Sakaguchi's experiment was scientific: he showed that among the immune cells of a healthy mouse, there must be some that stop immune reactions and can stop autoimmune disease.

Every moment in history has its own history. The idea that some types of immune cell might be there to stop, rather than start, an immune response had come up before. Through the 1960s and early 1970s, the quest to understand our immune system was driven by methods which allowed different types of immune cells to be separated from one another. These methods were crude by today's standards but at the time, different types of immune cell could be separated and mixed back together to test how different combinations responded to germs or molecules from germs. This led to discoveries about how immune cells help each other and, as we discussed in Chapter Two, led to the discovery of the dendritic cell as being important in kickstarting a reaction. In the early 1970s, several research groups around the world found that the addition of some types of immune cell would supress, not boost, a reaction.[11] Richard Gershon, working at Yale University with his assistant Kazunari Kondo, published their observation of this in the British journal *Immunology*, after being advised by a colleague that this was one journal 'inclined to consider unorthodox data'.[12]

The idea that some cells could stop an immune reaction was controversial from the get-go. Especially problematic was that Gershon found that the cells able to do this were T cells, the very same type of immune cell already well established as being able to boost immune responses. Gershon proposed that there had to be some T cells that behave differently to normal T cells and he coined the term 'suppressor T cells' to describe those that could stop, rather than help, an immune response.[13] A decade later, Sakaguchi's experiment helped vindicate Gershon's idea – and extended it, showing that suppressive immune cells could be especially important in preventing autoimmune disease. But Sakaguchi stopped short of using Gershon's term, suppressor T cells. He referred to them as autoimmune-preventing cells, because he wasn't sure if the cells he used were the same as those that Gershon had described.[14] They could have met to discuss it, but sadly never did. Soon after Sakaguchi's results were published, Gershon should have been enjoying a peak in his scientific career but his life was cut short by lung cancer. He died aged fifty, when his daughter, Alexandra, was one.[15]

Gershon's obituary in the *New York Times* likened the discovery of another side of the immune system to seeing the other side of the moon.[16] Gershon, Sakaguchi and others were celebrated for a while because the idea that the immune system needed something to stop its activity was a powerful logic.[17] But their work wasn't taken to be definitive because there were other ways to explain what they observed. Another way of interpreting Sakaguchi's experiment, for example, was that changes to the immune system caused by surgical removal of the thymus might, say, allow a virus to prosper.[18] If symptoms were not due to an autoimmune disease but instead caused by a virus, then it would not be surprising that an inoculation of T cells from a healthy mouse would help stop the problem by fighting the virus. Sakaguchi felt confident that this wasn't happening, but he couldn't formally exclude the possibility.[19]

The main obstacle that prevented problems such as this from being quashed was the inability of anyone being able to separate suppressor T cells from normal T cells. The methods available were simply too coarse. Recall how the discovery of dendritic cells, discussed in Chapter Two, was only widely accepted after the cells could be isolated and then shown to have properties above and beyond other types of immune cell. Without a way to identify and isolate suppressor T cells, it was hard to prove their existence, let alone understand how they might work. But that didn't stop scientists guessing.

All kinds of ideas were suggested as to how suppressor T cells might work – about how different types of T cell might interact with each other or how antibodies might stick to each other. In hindsight, this period, from the mid-1970s to mid-1980s, was something of a Dark Age in studying the immune system, because while there was an abundance of complicated ideas about how the system might work, there weren't the tools to identify and manipulate genes and proteins that were required to test them. Theories built up like houses of cards. New words were invented – idiotypes, epitypes and paratopes – which nobody uses today. It's hard to follow many of the papers written during these Dark Ages. As L. P. Hartley famously wrote, 'The past is a foreign country: they do things differently there.'[20]

Eventually, new methods allowed for greater rigour – and suppressor T cells were caught up in the cull of ideas that ensued. One especially damaging episode was when, in 1983, a region of the genome thought to control the function of suppressor T cells was shown to lack any such gene.[21] Belief in suppressor T cells collapsed. Labs immersed in studying these cells found it hard to maintain their funding.[22] Suppression became a dirty word: the topic became synonymous with bad science and an over-interpretation of scant data.[23] 'No realm of immunology has less credibility than that of suppressor T cells,' scientists wrote in 1992.[24]

This surely adds to the magnitude of the achievement of the small band of researchers who persisted. I once asked Sakaguchi what kept him going, where did his inner confidence come from? He replied, matter-of-factly, that he just wasn't convinced that the cells he studied were the same type of cells that others called suppressor T cells. Many of the characteristics ascribed to suppressor T cells didn't seem to fit the cells his lab studied, which meant that he wasn't put off by the missing-gene episode of 1983. Sakaguchi's confidence didn't come from arrogance, ego or a calming mentor, it came from the data he had in his lab.

Once again, it was new technology that moved things forward, as is so often the case. Tools were developed that could mark out different types of T cell with far more precision, tagging them according to the different molecules they had at their surface. When new technology opens a door, it is common for more than one lab to discover what's on the other side at the same time. In this case, it led to an important experiment carried out in 1993 by two different teams working independently.

Fiona Powrie had been following, in her own words, a 'global jet-setting vision' of becoming an accountant when she decided that medical research was her true calling. Her change of mind was motivated in part by her mother dying from an autoimmune disease, lupus.[25] For her PhD at Oxford University, working in the British immunologist Don Mason's lab, she discovered that rats which had some of their T cells removed would develop an autoimmune disease.[26] After her PhD, she moved to Palo Alto, California, to work at an institute owned by US-based company Schering-Plough, and there decided that the project she had been told to undertake wasn't as exciting as pursuing her PhD research further. Following her own plan, she decided to test whether or not mice would also develop an autoimmune disease if some of their T cells were removed.[27] Unknown to her at the time, researchers at another company, Immunex, had had the same idea.

The key to the advance that both teams made was to separate mouse T cells into two types. One group of T cells – formally called naïve T cells – was composed of cells that were ready and able to mount a defence should their receptor prove compatible with a new threat but which had yet to encounter such a germ and be deployed. The second set of T cells was composed of those that had already been 'switched on' and used in the body. This second set included a hotchpotch of T cells with different jobs, including those T cells that remain after the infection has been cleared in order to provide stronger immunity should the same germ attack again, as well as suppressor T cells, activated by the body's own components. The researchers transferred each of the two groups of immune cells into a different set of mice, all of which had been genetically engineered to lack their own T cells, so that the only T cells present in the mouse would be those infused into it.

They found that in the first group, the naïve T cells, those which had never been switched on before, would, in the absence of suppressor T cells, switch on and attack the mouse's healthy tissue, so that they developed an autoimmune inflammation in their gut. This established, albeit in this unnatural situation, that normal T cells could attack healthy tissue and cause an autoimmune disease. If these same mice were then given a dose of the second group of T cells, the autoimmune disease was stopped.[28] This fitted precisely with the idea that those T cells responsible for fighting germs were also capable of attacking the body, causing autoimmune disease, but that other T cells – the suppressor T cells – could prevent this. The fact that two US research teams published their results within months of each other immediately validated the discovery.[29]

Meanwhile in Japan, Sakaguchi discovered a more precise way of identifying suppressor T cells. Instead of grouping cells according to whether or not they had been switched on before, he found, in 1995, that suppressor T cells had especially high levels of a particular cytokine receptor protein at their surface.[30] He used this information to remove this set of T cells from the mouse immune system.[31] To do this, T cells were taken from one mouse and those with the particular receptor protein were killed off. The remaining T cells were then injected into a second mouse, which again had been engineered to lack its own T cells. This second mouse now suffered autoimmune disease. This meant that removing suppressor T cells from a mouse's immune system was sufficient to cause illness. This directly supported Sakaguchi's big idea: that an abnormality in suppressor T cells could be what underlies many different types of autoimmune disease.

Ethan Shevach at the US National Institutes of Health read Sakaguchi's publication of this experiment and wasn't sure what to make of it. He had been a vigorous opponent of the idea of suppressor T cells, but on the face of it this experiment was a striking breakthrough. He charged a new arrival in his lab, Angela Thornton, with the task of repeating Sakaguchi's work. A thumbs up or down from Shevach's lab would be an important signal to the mainstream immunology community whether Sakaguchi's science should be celebrated or condemned.

Thornton found everything Sakaguchi had done to be true. And with that, Shevach changed his mind about the existence – and vital importance – of suppressor T cells.[32] So convinced of their importance was he that he shifted the bulk of his lab's research efforts to studying these cells.[33] Shevach was hugely respected – he was, at the time, the editor-in-chief of the *Journal of Immunology*[34] – and well known to argue aggressively against the idea of suppressor T cells, so the fact that he changed his mind about them turned heads everywhere. Sakaguchi recalls that many more people, especially in the USA, took greater notice of his work after Shevach's conversion.[35]

Shevach and Sakaguchi both showed, in 1998, that suppressor T cells could also suppress immune responses in a lab culture dish.[36] These experiments were simpler and easier to interpret compared to those done in living animals, and they helped convince yet more scientists that suppressor T cells exist. But there was still a problem. All of the research so far had been conducted on animals, or with animal cells, and none of it had been shown to be true in humans. This was probably for the simple reason that so few labs were working on suppressor T cells, a result of there being such scepticism about their existence for so long.[37] Eventually, in 2001 – three decades after the idea of suppressor T cells had first been suggested – six different teams identified human suppressor T cells all at once.[38]

In hindsight, it seems strange that suppressor T cells were lost in the aether for so long. When some of the predictions about suppressor T cells turned out to be wrong, it seems that the scientific community threw the baby out with the bathwater.[39] The chief problem was that methods weren't available to isolate suppressor T cells so that they could be studied in detail. But my own view is that another issue contributed to the error: scientists of the era had been too quick to judge. The complexity of the immune system means that we now know that we cannot expect every interpretation of every experiment to be correct. We have become more aware of our 'limited capacity to get everything right in every report', as Ron Germain, one of Shevach's colleagues at the National Institutes of Health, has put it.[40] As our scientific knowledge matures, the social psychology of the scientific community matures with it.

By the time human suppressor T cells were widely accepted to exist, the name 'suppressor T cell' had already been used as a synonym for bad science for over a decade. It had to be changed; a new name for a fresh start. From here on, these cells were to be called regulatory T cells, or Tregs (said as T-regs, like the dinosaur T-rex).[41] After decades of observing the shadows of regulatory T cells, they were finally in the spotlight and accepted as a crucial part of our immune system: guardians of the galaxy.[42]

*

The next leap forward in our understanding of regulatory T cells and autoimmune disease has its roots in a very unlikely source: the Manhattan Project. In response to the Manhattan Project's successful production of the world's first nuclear weapons, the Mammalian Genetics Laboratory was set up in 1947 at the Oak Ridge National Laboratory in order to understand the hazards of radiation. This grew to become an enormous research endeavour which lasted almost sixty years. At its peak, Building 9210, known as the Mouse House, used sixty-six rooms to hold a total of 36,000 cages, each holding between one and six same-sex adult mice.[43] At one point, the head of the institute, Bill Russell, stacked cages of mice in an old Ford and drove to an atomic test site in Nevada. The mice were left in exposure chambers in the desert while a bomb was tested.

He then brought the radioactive mice back to Oak Ridge, where the effects of their genetic mutations were analysed.[44] Descendants of these mice, and those exposed to other types of radiation or mutagens, were used in countless experiments across multitudes of labs. You'll have your own view of whether or not this was inhumane, necessary or both. But it's what happened, and it informed our understanding of the risks of radiation as well as becoming a source of mice that could be studied as models of human genetic disorders.

One of the most famous mouse colonies at Oak Ridge was born in 1949. These mice had not been treated with any radiation or mutagen but, just by chance, out of all the animals bred at the facility, these mice happened to be born with something evidently wrong. Organs where immune cells congregate were enlarged and the mice died prematurely. In 1991, it was discovered that the problem with these mice was that they suffered a vigorous autoimmune disease.[45] This was a time before genetic analysis was easy and it took another six years for a rough region of the genome to be identified as being mutated in these mice. There were twenty different genes in this region and the last of these to be tested individually turned out to be the single gene which had been altered to give these mice autoimmune disease.[46] It was a gene named forkhead box P3, known as Foxp3 (said Fox-P-3; the cumbersome name coming from a related gene first studied in fruit flies whose mutation results in the insect having a fork-headed appearance). A small fragment of DNA had been inserted into this gene by chance, preventing the gene from working properly and causing autoimmune disease.[47]

Such a dramatic effect in animals meant it was likely that mutations in the same gene in humans would also lead to illness, and it didn't take long for this to be proven right. Mutations in the human Foxp3 gene were identified in patients with a rare syndrome called IPEX (which stands for immune dysregulation, polyendocrinopathy, enteropathy, X-linked syndrome).[48] This syndrome – so rare that its prevalence is not known – is characterised by an overwhelming autoimmune attack on several organs.

The crucial question was why? What was the role of the Foxp3 gene, such that its malfunction caused autoimmune disease? A clue lay in the fact that the symptoms of IPEX syndrome were similar to the autoimmune disease that developed in mice when their regulatory T cells had been removed.[49] This led to the idea that perhaps the Foxp3 gene and regulatory T cells were somehow connected.

In 2003, three research teams – Sakaguchi in Japan and two research teams in the US, led by Alexander Rudensky and Fred Ramsdell[50] – discovered that the activity of the Foxp3 gene is not only linked to regulatory T cells, it is essential for their development and functioning.[51] In fact, the activity of this one gene has the power to change a normal T cell into a regulatory T cell, transforming a cell's purpose from boosting to dampening an immune response. This in itself was a dramatic discovery: that a single gene, switched on or off, can change a cell's core nature. The reason that this one gene, Foxp3, is so powerful was found to be that it encodes for a protein that directly controls the activity of around 700 other genes.[52] It is a hub in the network, a master control gene.

With this discovery, research into regulatory T cells exploded. Foxp3 was a far more reliable marker of these cells than anything used previously, and this allowed regulatory T cells to be tracked, isolated and systematically studied. The research that followed revealed that regulatory T cells safeguard against unwanted immune responses in more than one way. They secrete cytokines that dampen immune responses locally, and they can switch off the activity of another immune cell with a touch. One place in the body where it turned out that regulatory T cells are especially abundant is the gut. Here the immune system must be especially adept at knowing what's harmful and what's harmless, to distinguish 'salmon from salmonella', as Powrie puts it.[53]

In addition, regulatory T cells in the gut have what is probably the hardest job in the immune system. Commonly, the immune system should react against bacteria found inside the body, but in the gut, regulatory cells have the task of preventing any adverse reaction to the bacteria that live there to our benefit, the gut microbiome. These bacteria help digest plant molecules that are otherwise indigestible, extract nutrients and synthesise vitamins, all in return for a place to live. This is a symbiotic relationship that our immune system must preserve, not react against.

In fact, the immune system does more than just preserve this relationship, it shapes it. The gut is complex in the extreme: there are trillions of bacteria there – as many as there are human cells in the whole body[54] – living, dying, competing and cooperating. Plus there are untold numbers of viruses and fungi in our gut, of which we know very little. It is an inner universe that varies from person to person and that changes during puberty, pregnancy and seemingly every other physiological and pathological state.[55] Our gut microbiome can even change each time we eat or defecate, and the immune system must be able to tolerate this flux while still being able to mount a robust defence when needed. If the immune system misses a threat, we are left open to the array of illnesses which can be caused by germs in our food or drink; equally dangerously, if the immune system overreacts to our resident gut bacteria, the resulting inflammation may cause anything from mild discomfort to chronic bowel disease.

In order to adjust its behaviour appropriately and maintain different bacteria in the gut as necessary, the immune system switches on and off in response to small molecules called metabolites, which are by-products from the replication and growth of gut bacteria.[56] Metabolites from desired bacteria dampen the sensitivity of immune cells, counteracting their tendency to switch on in the presence of bacteria.[57] Likewise, if metabolite levels from favoured bacteria fall, then the immune system takes this as a cue that unwanted, potentially harmful, bacteria have begun to displace the normal healthy flora. The immune system then kicks into action to defend us *and* our resident gut bacteria. In this way, our immune system does more than protect us from disease; it directly maintains the vital symbiosis between us and the bacteria that colonise us.

The gut immune system also looks out for trouble more directly – by sensing molecules that in any normal situation would operate inside a cell and whose presence in the gut alerts the system to the fact that cells have been burst open, for example when bacteria or viruses leave a cell. Such molecules usually have nothing to do with the immune system when they are inside the cell – they might be important for cells to replicate or to move[58] – but once outside they act as a bat-signal that there's a problem, and are referred to as alarmins.

These discoveries fit with a big idea that had been proposed by Polly Matzinger, chief of the 'T Cell Tolerance and Memory Section' at the US National Institutes of Health (NIH) almost a decade before the link between Foxp3 and regulatory T cells was established. Matzinger had thought deeply about Janeway's suggestion, in 1989, that the immune system can't work solely by detecting things alien to the body but must specifically detect germs. She realised that even this needn't be the case: the body doesn't need to trigger an immune response in response to any virus or germ; it needs only to respond to germs that cause damage. An effective immune system, Matzinger concluded, only needs to defend against things that are dangerous,[59] and she proposed the overarching principle that the immune system works by sensing damage to the body.[60]

Her publication of this idea in 1994 caused a riot.[61] One side claimed that her thesis was as revolutionary as the Copernican revolution of the sixteenth century, which removed the earth from being the centre of the universe. Others retorted that 'it is unclear how the inferences and conclusions drawn by the authors could have passed peer review'.[62] Some of this resistance might have been because Matzinger came to science with a less-than-common background, having worked as a jazz musician and as a Playboy Bunny at a club in Denver.[63] And she had a reputation for being mischievous: one of her research papers, published in one of the world's best scientific journals, included the name of her pet dog as co-author.[64] As a result, she was banned from publication in that journal for the next fifteen years – until the editor died, she has suggested. When the NIH initially considered her for its permanent faculty, they concluded that the dog co-authorship issue wasn't fraud because the dog had visited the lab and had done no less research than some co-authors on other papers.

Today, Matzinger's idea is far less controversial: there is plenty of evidence that immune responses in the gut and elsewhere are driven and shaped by damaged tissue. My own view is this doesn't mean her idea should directly supersede other ideas about how the immune system works. Rather, we must not expect everything the immune system does to fit any one over-arching principle. The system discriminates between self and non-self, and it detects germs, and it responds to danger, and it does all these things concurrently – and messily. The immune system uses a collection of mechanisms which no single principle fully encapsulates.[65]

As an example of how complex things are: one type of alarmin that is released when the lining of the gut is damaged switches on regulatory T cells rather than normal T cells. This turns off the immune system rather than switching it on.[66] Although damage indicates a problem that may very well be due to an infection that warrants an immune response, restraint is also needed to prevent the immune system from spiralling out of control and causing more damage. The level at which this alarmin dampens an immune response is altered by the levels of other molecules present, including cytokines that are them-selves signifiers of levels of invading germs.[67] An inner universe of small molecules, of metabolites, alarmins and cytokines, reflecting the presence of different gut bacteria, invading germs or damaged cells, all dial the activity of the immune system up and down.

This complex blend of triggers and restraints is also tuned by the food we eat. The best-known job that our gut bacteria perform is helping to digest fibre from fruits, vegetables, or cereal grains, which our body would otherwise struggle with. Diets high in fibre have a wide range of overall effects on the body, from reducing blood pressure to lowering the risk of colon cancer.[68] They also affect our immune system specifically; many of the molecules produced when bacteria break down soluble fibre stimulate the production of regulatory T cells.[69] At least in mice, a high fibre diet increases the number of regulatory T cells, which helps protect against autoimmune disease.[70]

To test the relationship between the microbiome and the immune system, methods have been developed for producing mice with drastically depleted microbiomes, which involves heavy doses of antibiotics, and with no microbiome at all, which involves breeding them in a facility that is devoid of all microbes. The latter enjoy the rare distinction of having a scientific name that also works in plain English: germ-free mice. They are born and kept all their lives in sealed-off see-through plastic enclosures and are fed through portals food which has been irradiated to kill germs. They are probably the only creatures on earth not colonised by something else. In both situations, when a mouse's gut microbiome is either depleted or removed, their immune system alters dramatically, including an enormous reduction in regulatory T cell numbers.[71] Nonetheless, mice can live a long life in germ-free facilities, so it's not that their microbiome is absolutely essential. It's that mice, humans and all other animals have evolved to live in a microbe-filled world, just as we've evolved to live on a planet with a twenty-four-hour cycle of light and dark, and when this changes, parts of us, perhaps our immune system especially, are disorientated.

It's possible that the average human microbiome has altered since the advent of modern hygiene, now that we are exposed to far fewer germs than our species would have been accustomed to in centuries past. It may be, for example, that it is less diversely populated than it once was, reducing the number of regulatory T cells we have. Having fewer regulatory T cells would lead to less restraint on the immune system, which could feasibly account for the rise of all kinds of allergies, including food allergies, as well as autoimmune diseases. This fits with the 'hygiene hypothesis' first proposed by David Strachan, working at St George's Hospital in London. By studying a survey of over 17,000 children born in March 1958, he calculated, in 1989, that whether or not they had ended up with allergic hay fever correlated with the size of the family into which they had been born and especially with how many older siblings they had.[72] He realised that, on average, infections would occur less frequently in smaller families. This led him to suggest that hay fever might be prevented by the contraction of infections early in childhood. In turn, this led him to suggest that, more broadly, allergies may become more commonplace with increased hygiene. His idea has guided our thinking about allergies ever since.

Of course, the hygiene hypothesis doesn't imply that modern standards of hygiene are bad for our health. After all, they have brought about a dramatic fall in infectious disease and few experts, if any, would think it wise to bring back infections or parasites that we have slowed down or even eradicated. The hypothesis also doesn't mean that we should wash less often; there is no evidence that frequent showering or bathing increases the risk of allergies or autoimmunity. On the other hand, there is evidence that children growing up on small farms are less likely to develop allergies. So something about a 'dirty' environment may help, and the important question is what, exactly?

To answer this, scientists have studied two relatively isolated farming communities living in the USA: the Amish and the Hutterites, who have similar ancestry but differ in their suscep-tibility to asthma. Amish children have asthma relatively rarely – around 5% are affected – while the prevalence of asthma among Hutterite children is around four times greater. Both communi-ties have large families, similar diets and get childhood vaccin-ations, but one difference is that the Amish use traditional farming methods on single-family dairy farms while the Hutterites embrace larger-scale communal mechanised farming. Each community lives in a similar environment but Amish chil-dren live closer to animals and their sheds.[73] The fact that the Amish are less likely to get asthma corresponds with the hygiene hypothesis: stimulation of the immune system by microbes found on small farms might be what protects the Amish from asthma.

To test whether or not there were any differences in the state of the immune system in Amish and Hutterite children, a team of researchers, primarily from the universities of Chicago and Arizona, analysed the numbers of different immune cells and which genes were active in blood samples taken from sixty schoolchildren.[74] They found that the Amish children's innate immune cells, which recognise the telltale signs of germs, were being continuously stimulated at a low level. In other words, their immune system was continuously being tickled by the presence of bacteria.

To test more directly whether or not microbes from Amish farms could impact asthma, mice with symptoms of asthma were given doses of microbe-carrying dust from Amish and Hutterite homes. Amazingly – a 'gee-whiz moment', one of the research team put it – dust from Amish, but not Hutterite, homes could suppress symptoms of asthma in mice.[75] Fitting with the hygiene hypothesis, this showed that microbes from Amish farms could help tackle asthma. This dramatic result doesn't, unfortunately, lead to an immediate medical cure. 'We could just say put a cow in everyone's house and no one would have asthma anymore but of course that's not easy,' one of the research team half-joked to the *Washington Post*.[76] But as our understanding deepens, we may uncover new ways to nudge the immune system to stop asthma, perhaps using inactivated versions of the bacteria present on dust from Amish farms.

The use of antibiotics has also been linked with increasing the risk of allergies. Antibiotics are of course hugely important for our survival against bacterial infections, but they have been overused – given, for example, to people with viral infections, against which they are absolutely ineffective. Even people who think they haven't used antibiotics for some time may well have taken them unwittingly because antibiotics used in farming can, in some places, seep into our food and water.[77] One well-known consequence is the increased prevalence of drug-resistant bacteria. Because of this, health organisations around the world are trying to use antibiotics less often. Far less discussed, however, is the possibility that antibiotics also damage our resident gut microbes and change a person's microbiome.[78] The use of antibiotics by children, or mothers during pregnancy, has been linked with childhood asthma but this does not in itself prove that using antibiotics increases the risk of asthma; the correlation is very likely caused by genetic or environmental factors which link families to both asthma and infections that require antibiotics.[79] The consequences, if any, of antibiotics changing a person's microbiome remain unclear.

Antibiotics are far from being the only factor known to influence our gut microbes; another is where we live. One study has compared the microbiome of children in Finland, Estonia and Russia. Childhood autoimmune disease is relatively common in Finland and Estonia but far less likely in Russia. 222 children – seventy-four from each country – had their stools analysed every month for three years to profile the bacterial contents of their microbiome. In parallel, parents filled out questionnaires about breastfeeding, diet, allergies, infections, family history and drug use.[80] This huge undertaking revealed that geographical location was a major factor in an infant's microbiome. Independent of diet, the use of antibiotics and any other confounding factor, particular types of bacteria were especially high in Finnish and Estonian children, while others were more common in Russians, especially during the first year or two of life.

What's more, while a molecular component of bacteria dominant in the microbiome of Finnish and Estonian children is known to switch off our immune cells, the equivalent molecule in bacteria common in Russian children, which has slight differences, tends to have the opposite effect: it switches on immune responses. This fits with the idea that the make-up of a child's gut bacteria can impact how their immune system develops – and that bacteria common in Russian children may help protect against autoimmunity, because switching on an immune response early in life helps train the system to respond appropriately later in life.[81] Again, this important scientific advance doesn't quite lead us to an easy medical solution for allergy or autoimmunity. We can't deliberately expose healthy children to a concoction of germs, or molecules from germs, when the side effects are unclear and the risks of them developing allergies or autoimmunity are also unclear. But one acceptable intervention might be instead to control or supplement the food we eat.

Vegetable fibre or supplements which encourage gut bacteria to multiply – so-called prebiotics – could feasibly nudge the state of our immune system to our benefit, but it's difficult when nurturing one kind of bacteria to ensure that a closely related but detrimental species of bacteria doesn't also thrive.[82] Another idea is to ingest live bacteria, in yoghurt or other foods – so-called probiotics – which could also feasibly shift the make-up of our gut microbiome and in turn impact the state of our immune system. There's no unambiguous evidence that either prebiotics or probiotics help, but the design of food supplements, which are not currently evaluated in the same way that medicines are, will undoubtedly become more sophisticated, more precise and more medical, as our understanding improves.[83]

One way that probiotics could become more sophisticated would be to use genetically modified live bacteria. Such upgraded bacteria are technically easy to produce; it's a standard recipe that any biology lab can follow – mix bacteria with a new gene, add some chemicals, give a pulse of electricity and you're done. Even inserting human genes is easy: bacteria were modified in this way to produce insulin as long ago as 1978. In that case, pure insulin was used as medicine, the bacteria only being used in the industrial process of producing it, but essentially this is the same technology needed to create genetically modified live bacteria which could be added to food directly. In mice, bacteria engineered to produce a cytokine which normally comes from regulatory T cells can stop the symptoms of autoimmune disease.[84] This has not yet been achieved in human clinical trials, but new medicines like this, and others we haven't yet conceived, will emerge as our understanding of regulatory T cells increases. And this is just the tip of an iceberg.

Sakaguchi was right all along to think that finding out what can cause the body to attack itself would lead to a deeper understanding of how the immune system works. Before him, the dogma was that immune cells capable of reacting against the body's own components were weeded out from the system, killed off in the thymus without ever reaching the bloodstream. But Sakaguchi and his contemporaries revealed the situation to be more complex than this. The system specifically includes cells able to detect the body's own components, which are there to safeguard against an immune reaction. We now know that this was just the tip of an iceberg because in fact, there are many types of T cell; far greater diversity than can be covered by the crude categories of 'normal' or 'regulatory' cells.

Indeed, the classification of immune cells as a whole has been too coarse. They have been put into boxes according to a few identifying marks – usually the presence of one or two specialist protein molecules – and a broad understanding of what they are capable of: white blood cells called Natural Killer cells are good at killing cancer cells, macrophages are good at gobbling up bacteria, and so on. We now realise that there are many different types of Natural Killer cells, many different types of macrophages; that each of the categories of cell we have invented contains many sub-categories. In fact, one study has found a way to classify thousands of different types of Natural Killer cell.[85] As we have just seen, within the same category of immune cell, some work to switch on an immune response, while others switch it off. What's more, each flavour of immune cell behaves differently depending where in the body it is. Immune cells in the gut tolerate bacteria more readily than in the lung. Bluntly, it's hard to understand how the system achieves all that it does. On the other hand, perhaps we shouldn't be so surprised; it's hard, if not impossible, for any one person to understand something as seemingly simple as the Google search engine: scanning the Internet for a few words is a part of daily life, but behind the scenes it involves a hugely complex array of algorithms, each designed by a team whose expertise is unlikely to extend beyond a small part of the whole.

The reason that we've begun to triumph – why it is not hyperbole to suggest that we are at the dawn of a health revolution – is that we have now identified some of the hubs in the system: cells and molecules that, when targeted with drugs that boost or halt their activity, dramatically shift the behaviour of the system as a whole. We saw this with anti-cytokines. Blocking only one cytokine, TNF, for example, can alleviate the inflammation that underlies arthritis by halting an entire cascade of effects – in this case by severing the feedback loop in which immune cells keep triggering one another into action, leading to an autoimmune attack. When drugs, foods, prebiotics or probiotics are developed to impact the behaviour or numbers of regulatory T cells, which are undoubtedly also a hub in the system, we will have new treatments for allergies and other autoimmune diseases.

In the meantime, where this revolution will strike next, and to some extent has already struck, is in the treatment of cancer. There, a new branch of medicine has recently been opened up by the discovery and understanding of a whole other way in which the immune system is regulated.

8 Future Medicines

'Every time Jim meets a patient, he cries,' Padmanee said to the *New York Times* in 2016.[1] 'Well not every time,' Jim added. Jim Allison and Padmanee Sharma work together at the MD Anderson Cancer Center in Houston, Texas, having met in 2005 and married in 2014. A decade before they met, Allison and his lab team made a seminal discovery that led to a revolution in cancer medicine. The hype is deserved; cancer physicians agree that Allison's idea is a game-changer, and it now sits alongside surgery, radiation and chemotherapy as a mainstream option for the treatment of some types of cancer.[2]

Take one example. In 2004, twenty-two-year-old Sharon Belvin was diagnosed with stage IV melanoma – skin cancer that had already spread to her lungs – and was given a 50/50 chance of surviving the next six months. Chemotherapy didn't work for her and her prospects looked bleak. 'I've never felt that kind of pain,' she later recalled, ' … you are lost, I mean you're lost, you're absolutely out of control, lost.'[3] All other options exhausted, she signed up to an experimental clinical trial testing a new drug based on Allison's idea. After just four injections over three months the tumour in her left lung shrunk by over 60%. Over the next few months, her tumours kept shrinking and eventually, after two and a half years of living with an intense fear of dying, she was told that she was in remission – her cancer could no longer be detected. The treatment doesn't work for everyone but, Allison says, 'We're going to cure certain types of cancers. We've got a shot at it now.'[4]

Once she had recovered, Sharon Belvin was the first patient Allison met.[5] Her parents and husband were also there – and everybody was crying. Belvin hugged Allison tight. 'There's no words to describe what it feels like,' she has said, ' … what it feels like when you have handed someone back their life.'[6] Both the Talmud and the Koran teach that when a person saves a life, it is as if they have saved a whole world. About two years after they met, Belvin sent Allison a photo of her first baby, and a couple of years after that, she sent him a photo of her second.

Not just a one-off success, this major new medicine has saved or prolonged thousands of lives. But it came about not from any attempt to treat a particular type of cancer, or any disease for that matter. Rather, we owe its existence to curiosity-driven research – a tinkering with cells and molecules – to fathom how the immune system works, and we're only just beginning to understand its potential benefits.

Cancer was once thought to be invisible to our body's defences. As cancer is rarely caused by a germ and is rather an abnormal expansion of our body's own cells, there tends not to be anything as obvious as a molecule from a virus, bacteria or fungus to mark out a cell as cancerous, and for a long time a view was widely held that cancer displayed nothing to the immune system that it could recognise as alien. As long ago as 1943, the first evidence that the immune system can respond to cancer (the type not caused by a virus) was published,[7] but the idea remained controversial for over three decades.[8] This was because of the possibility that the immune reactions observed in the experiments in question had been caused not by the tumours in the animal subjects but by the chemicals used to induce the tumours. Eventually, several lines of evidence established that our immune system can and does fight cancer: immune cells were found to infiltrate tumours, and when isolated in the lab, these cells could kill tumour cells. In addition, mice engineered to lack a proper immune system were found to be especially susceptible to cancer.[9]

Work by Belgian scientist Thierry Boon, among others, established definitively that the genetic and epigenetic changes that turn a cell cancerous are sufficient for it to be detected by the immune system.[10] Boon identified fragments of proteins that had been altered in cancer cells which can be detected by T cells as not having been in the body before. The implication of this discovery is that, as well as seeking invading germs, the immune system helps maintain the integrity of our own body's cells, screening against detrimental genetic mutations that can arise whenever cells divide.

All immune responses are multilayered and our body's defence against cancer is no exception. As well as T cells, the white blood cells known as Natural Killer cells are also able to fight cancer, as we've mentioned already. Like T cells, they do so by sending a packet of toxic proteins into a cancerous cell, but they have different strategies for detecting when a cell has turned cancerous, one of which involves recognising protein molecules not normally found on healthy cells but which cancer cells sometimes display at their surface – the so-called 'stress-inducible proteins' we met in Chapter Five. (Nonetheless, it is likely that cancer is recognised by our immune system less easily than is, say, the flu virus, though this is hard to prove.)

The discovery that our immune system can fight cancer led in turn to the possibility that we might tackle the disease more effectively by harnessing or boosting our immune response. In fact, this type of therapy, often called immunotherapy, has a long history. Carried out well before T cells or Natural Killer cells were even known about, a series of well-documented experiments by William Coley, performed through the 1890s, are often credited with the birth of immunotherapy. Coley, a surgeon at the Memorial Hospital, New York, noticed that a patient with neck cancer began to improve when she had a severe skin infection. He then unearthed forty-seven similar cases in the medical literature which propelled him to systematically test whether or not a deliberate inoculation with a mixture of heat-killed bacteria – known as 'Coley's toxins' – could help cancer patients. In an era before institutional review boards, a twenty-nine-year-old surgeon could conduct a human trial essentially on a hunch.

A century later, as we will see, Allison would need to approach human trials very differently.

While Coley's toxins were successful in some patients, the overall effectiveness of the method was inconsistent, especially when other physicians tried to replicate it, because the concoction itself varied.[11] The medical establishment never embraced Coley's approach and explained away his successes as being down to initial misdiagnoses.[12] What exactly Coley used is no longer clear, and the charity Cancer Research UK has concluded that the scientific evidence does not show that Coley's toxins can treat or prevent cancer.[13] Still, something of what Coley attempted lives on. As described in Chapter Three, Steven Rosenberg found that cytokines which boost immune responses can sometimes help a person fight off cancer, although the problem with this is that cytokines switch on all kinds of processes and the storm of immune activity that follows can be toxic, occasionally lethal. So if there's one word that encapsulates everyone's idea as to what is most important when harnessing the immune system to fight cancer, it's *precision* – in terms of selecting only those patients for treatment who are actually predisposed to respond to it (something we'll come back to), and, most importantly for understanding Allison's success, in terms of boosting only the precise set of immune cells that will target a patient's cancer.

While scientists are all agreed that switching on the right set of immune cells is important, the difficulty lies in working out how to do it. One way of manipulating the immune response with precision is to use antibodies – the most precise biological agents we know. A natural part of our immune system, antibodies circulate in our blood and their job is to lock onto germs or infected cells and either incapacitate them directly or tag them for destruction. Antibodies can be produced to lock onto almost anything; as we saw in Chapter Four, antibodies used to block a cytokine can treat some people's rheumatoid arthritis. Allison's breakthrough idea was to also use antibodies – but in a way that's very different to how they had been used before.

His starting point was the process by which an immune response ends. When T cells initially detect an infected or cancerous cell, they multiply. Over a few days, a few hundred T cells can become many millions, expanding the small fraction of immune cells with just the right receptor to recognise the diseased cell. But an expansion of immune cells obviously cannot go on for ever and after a while, in the course of a normal immune response, T cells, and other immune cells, must switch off so that the immune response winds down and the system returns to its normal resting state, usually after the threat has been cleared. Maybe, Allison thought, by stopping this 'switch off' signal, immune cells could be set free to attack cancer cells more effectively, for longer. Building upon the idea that anti-bodies can be used to block the activity of proteins, his idea was to find a way to block the receptor protein that normally puts the brakes on an immune cell's activity.

The reason that this was something of a paradigm shift was that while others had sought a way to turn *on* an immune response against cancer, Allison's idea was to switch something *off* : 'To unleash, not harness … the anti-tumour response,' as he puts it.[14] The great advantage of this approach is its precision: only those cells that have been switched on to attack the tumour would in turn have their brakes put on, so only these cells, not every immune cell in the body, would be unleashed by the intervention. This approach has become known as immune checkpoint therapy.

Allison never set out to study cancer – 'That wasn't it at all,' he says – he set out to understand how T cells work.[15] But cancer was on his mind because he had lost his mother, two uncles, and then later his brother, to cancer, and had seen first-hand the ravaging side effects that radiation and chemotherapy can have. He had graduated from school early, started college at age sixteen and knew already then that he wanted to be a scientist. At that time, T cells had just been discovered as a specific type of white blood cell.[16] It was by studying the role of various receptor proteins on the surface of T cells that Allison, along with others, stumbled upon the T cell's brake system.

It's often said that a journey of a thousand miles begins with a single step.[17] But pinpointing that first step in the journey towards any scientific idea is especially hard because every new idea is built upon the ideas that preceded it. Allison's discovery was no exception. As we saw in Chapter One, Charles Janeway realised that the presence of something that has never been in your body before could not be the sole trigger for an immune reaction; a second signal must be needed. For Janeway this was the detection of germs; for Polly Matzinger, as we saw in the previous chapter, it was the detection of something dangerous. In Chapter Two, we saw how Ralph Steinman discovered that dendritic cells are particularly good at detecting germs. When they do so, they signal to T cells that a germ is present – providing the second signal needed to kick-start an immune reaction – by displaying so-called co-stimulatory proteins at their surface. These proteins on the dendritic cells fit into receptor proteins on the surface of T cells like a key in a lock, essentially unlocking the T cell's potential.[18] It's here that Allison's journey can be said to have begun: with the identification of a second receptor protein on the surface of T cells that was uncannily similar – about 30% identical – to the one 'unlocked' by co-stimulatory proteins, but whose role in our immune system remained a mystery.

The mysterious receptor had been given an especially cumbersome name: cytotoxic T-lymphocyte-associated molecule 4, or CTLA-4, named for simply being the fourth in a series of molecules identified on T cells. (It is also the registration number of Allison's Porsche.[19] Presumably there's far less of a demand for the number plate 'CTLA4' than there is for, say, his first name 'Jim'.) Not that Allison discovered CTLA-4 himself: it was identified in 1987 in Pierre Golstein's lab in Marseilles, as part of his lab's mission at that time to discover which genes are active exclusively in T cells and no other white blood cells.[20] Golstein didn't pursue his discovery of CTLA-4 and uncover its role. He simply showed that it was present at the surface of T cells switched on to participate in an immune response – whereas it was not present on T cells that were resting, simply waiting for signs of trouble. This indicated that the molecule was somehow important only once an immune response had got going. Not much to go on, but intriguing.

Recounting what each person then did to unravel the mystery of CTLA-4 is not easy.[21] When, in 2015, Allison won a prestigious medical award for his work, an article in the *New York Times* commented that recognition of a single individual presents a false picture of how new medicines come about, on the basis that an analysis of prior research quoted in Allison's papers showed that his discovery built directly on the work of 7,000 other scientists at 5,700 institutions.[22] That doesn't even include the physicians and patients involved in the clinical application of Allison's idea, nor those in the pharmaceutical industry who turned a lab molecule into a licensed medicine. On the other hand, as one eminent immunologist wrote: 'It is rare that such a sea change can be traced to any one individual, but the advent of checkpoint therapy would have been highly unlikely without the efforts of James Allison.'[23] To me, both views seem right. It takes a village – and an individual spark – to raise a medicine.

The results of experiments aimed at revealing what CTLA-4 does in the body were, at first, interpreted in line with expectation: that the protein receptor helped stimulate T cells. It was, after all, very similar to another stimulatory receptor, and redundancy of this sort is built into the immune system, with many different molecules and cells having overlapping tasks, presumably to help the system be robust: such redundancies ensure that if a germ were to interfere with any one component, there's a good chance that its function can anyway be performed by another component. But in 1994, Jeff Bluestone and his research team at the University of Chicago – urged on by Bruce Springsteen's music, which was playing in the lab[24] – stumbled across the fact that what CTLA-4 seemed to do was the complete opposite to what was expected.

At the time, Bluestone's team had produced an antibody to block the CTLA-4 receptor (in the same way that Jan Vilček made an antibody against the TNF cytokine, discussed in

Chapter Four), which allowed them to test what would happen to T cells if CTLA-4 was incapacitated. His lab's overarching mission was to find ways of stopping an immune response in order to help address the problems of organ transplantation or autoimmune disease. Like everyone else, Bluestone presumed that CTLA-4 was likely to be a stimulatory receptor – an on-signal – and that by blocking it, the immune system would be rendered less effective.

He says he'll never forget the day when student Teresa Walunas came into his office and showed him the results: that blocking CTLA-4 with an antibody caused T cells to react more, not less.[25] If blocking CTLA-4 led to a stronger reaction, then CTLA-4 must normally deliver an off-signal, not a stimulatory signal. Because the result went so directly against the prevailing view, finding it didn't feel like a eureka moment. As Bluestone recalls, his reaction was more like: 'Gee, maybe it's an off-molecule but boy it will be hard to prove *that* to anybody.'[26]

Two of Bluestone's friends[27] had recently helped set up a new scientific journal called *Immunity*, so he published his discovery there. At the time, Bluestone was worried that his work might be hard for others to find if the new journal didn't take off, but he needn't have been concerned. It soon became one of the world's top journals for research about the immune system.[28]

Allison, then director of the Berkeley Cancer Research Laboratory at the University of California – and Bluestone's rival – had, in 1989, charged his own PhD student Matthew 'Max' Krummel with the very same problem: finding out what CTLA-4 does. Allison didn't have a specific idea he wanted to test; it was curiosity-driven rather than hypothesis-driven science, as Krummel puts it.[29] Initially unaware of what Bluestone's lab were up to, Krummel also made antibodies that could lock onto CTLA-4 so he too could test what happens to an immune response when CTLA-4 is blocked. Making an antibody wasn't so easy at that time – it is much easier nowadays – and it took Krummel four years to hit upon a recipe that worked.[30] Once he had the antibody in hand, his experiments using it gave the same result as Bluestone's: blocking CTLA-4 could boost an immune reaction, consistent with the idea that CTLA-4 normally delivers a switch-off signal to T cells.[31]

Even though both Bluestone's and Allison's labs came to this same conclusion, their discovery was still controversial. In part, this was because an antibody stuck to CTLA-4 may block a receptor from working, but it may also, in principle, do the opposite and trigger it, and blocking a switch-off signal would provide the same result as triggering a stimulatory signal.[32] The controversy was resolved when mice genetically altered to lack CTLA-4 were found to die at a young age because of a massive expansion of immune cells, overrunning the animal's body, producing toxic levels of inflammation.[33] This plainly showed that CTLA-4 was vitally important for switching off an immune response – and also established that switching off an immune response is just as vital to health as switching it on.

Next, Krummel produced large amounts of his antibody in order to test the effect of blocking CTLA-4 on different types of immune response. But the particular experiment which proved decisive and led to a medical breakthrough was one that Krummel didn't have time to do himself. Krummel was busy testing how blocking CTLA-4 affected immune responses against proteins from bacteria,[34] so Allison asked a new arrival in his lab, Dana Leach, to test how blocking CTLA-4 affected the immune response to tumours.

Leach injected the antibody into mice with bowel cancer. Allison hoped that by blocking the T cell switch-off signal, tumours in the colon might be attacked more effectively by the immune system and their growth might slow down. The results were even better than he had hoped. 'When Dana Leach … showed me the initial data, I was shocked and surprised,' Allison later recalled.[35] In all the animals treated, the tumours had regressed completely.[36] Over the Christmas holiday of 1994, they repeated the experiment blind, so that the person analysing the mice wouldn't know which animals had been given the treatment. Leach set up the experiment, went off to see his girlfriend for the holiday season, and Allison himself measured the tumours.[37] At first, the tumours stayed the same size. Then, after two weeks – 'as if by magic'[38] – the tumours began to regress in just one group of mice. A little later, they had completely disappeared. That group of mice was, of course, the one that had received the treatment. 'The fact that blockade of a single molecule could lead to complete tumour regression was astonishing,' Allison said.[39]

Over the next fifteen years, Allison's team, and others, found that blocking CTLA-4 could help treat many different types of cancer in mice. Once again, good news for mice but the next step was to test this in humans. In hindsight it is hard to believe, but Allison met with great resistance when he approached companies and funding agencies to do this. At the time many physicians, academics and industry scientists were profoundly sceptical about harnessing the immune system to fight cancer because so many previous attempts, using cytokines or dendritic-cell vaccines for example, had largely failed and led to complicated side effects. 'Some of my friends, very renowned,' Allison recalls, '[if] they wanted to insult me in public, they'd say Jim's a tumour immunologist, [and] snigger.'[40]

It took around two years before Allison could get a company interested in his idea.[41] Eventually, immunologist Alan Korman at Nexstar, a biotech company in Colorado, began working on an antibody to block human CTLA-4, after licensing the idea from Allison's university.[42] Nexstar sublicensed the idea to another company, Medarex, based in New Jersey, who had recently acquired a third company, GenPharm, which specialised in producing antibodies that could be safely used in humans. All this led to an antibody – named MDX-010 – which Allison and others could use in clinical trials. The process couldn't have been any further from William Coley's in the 1890s, who simply tested his idea on patients without delay.

In the first small trials, MDX-010 gave a durable response in a fraction of patients but led to adverse side effects in other patients. Larger trials gave mixed results. What helped save the drug from being dropped was that it performed much better when the criteria by which a cancer treatment was deemed successful were changed. That's not to say the rules were twisted simply to make the drug look better. Rather, astute clinicians realised that in some cases, according to the existing rules, this new medicine would be regarded as a failure even when patients actually benefited.

The reason was that success for cancer drugs had been defined with chemotherapy in mind. These agents often kill cancer cells directly and if the treatment is successful, a person's tumour can become smaller within weeks. For trials using the antibody to block CTLA-4 – unleashing the power of the immune system – little might happen at first. Measurements of the tumour would sometimes show that its size increased, formally indicating that the treatment had failed. But these numbers lied. Later, presumably after the immune system had been given enough time to get going, the tumour might then shrink.[43] We now know that a tumour might initially get bigger after treatment – seemingly bad news for the patient – because immune cells move into the tumour causing it to swell – which is actually good news for the patient.[44]

For medicines which harness the immune system, the World Health Organization's criteria for success in a cancer trial had to be changed. These rules – now known as the Immune-Related Response Criteria – included an increase in the time allowed for the treatment to work.[45] An initial increase in tumour size was recognised as a tumour flare, and not necessarily a sign that the treatment had failed. These changes transformed the fate of what would become a life-saving cancer drug and shows why the relationship between the pharmaceutical industry and regulators is complicated; they must work together, as well as independently, to scrutinise new medicines.[46]

When their interim data suggested that using their antibody to block CTLA-4 was not superior to chemotherapy, the pharmaceutical giant Pfizer dropped the idea, but Medarex persevered.[47] Pfizer may have just made the decision too soon; an improvement in the overall survival of patients might have become apparent later.[48] At any rate, Pfizer sold the rights for their antibody to another company, MedImmune, owned by AstraZeneca. Once the success of this type of therapy became clear, Pfizer changed its mind and agreed to pay up to $250 million to a small biotech company in 2016 in order to once again have proprietorship of an antibody able to block CTLA-4.[49] Medarex, on the other hand, were rewarded for their perseverance. In 2009, New York-based Bristol-Myers Squibb bought Medarex for over $2 billion, primarily for their antibody that blocked CTLA-4.[50] At the time, the antibody was still being assessed in clinical trials; it hadn't been proven to work but was evidently worth a $2 billion bet. Needless to say, a lot of deal-making takes place alongside the science when a new cancer drug is at stake.

Shortly after Medarex was acquired by Bristol-Myers Squibb, blocking CTLA-4 was proven to work, thanks to the new Immune-Related Response Criteria. In a decisive trial of melanoma patients, overall survival was used as the principal criteria for evaluation rather than other measures of a response such as a change in tumour size. For all the patients in this trial, their cancer had spread from the skin to other places in the body and life expectancy was short. The results, announced on 5 June 2010 to over 30,000 delegates at an annual cancer meeting in Chicago and simultaneously published in the prestigious *New England Journal of Medicine*,[51] showed that the average survival of patients went up from about six to ten months when they were treated with the antibody that blocked CTLA-4. This was an unprecedented result: no previous clinical trial had revealed anything to be capable of increasing the average lifespan of such late-stage melanoma patients. Even more amazing was that some patients enjoyed a long-term benefit. Over 20% lived another two years or more. Some of those who received the drug early on have since survived more than ten years.[52] Sharon Belvin is one.

In March 2011, the US Food and Drug Administration approved the new medicine, by which time it had been given a generic drug name ipilimumab (not necessarily a huge improvement on MDX-010) and the brand name Yervoy. Immediately, it was predicted that 15% of the 68,000 patients diagnosed with melanoma every year in the USA would receive

Yervoy. With a course of four doses initially costing upwards of $80,000, worldwide sales of the drug were predicted to reach $2 billion in 2015.[53] The hype turns out to have been almost justified: actual sales of Yervoy in 2015 were $1.1 billion.[54] Not surprisingly, the development of medicines to harness the immune system has become the fastest-growing area of the entire pharmaceutical industry.[55]

Much can be improved. There can be side effects from taking the brakes off T cells, including an unwanted inflammation in a person's skin, colon, liver or other organs. Some of this can be managed with counterbalancing drugs to suppress the immune system, but occasionally side effects can be life threatening. Perhaps the most challenging problem of all is that only a fraction of patients respond. Allison is all too aware of this: 'I wish it worked more.'[56]

★

Allison's wish – to help more than one in five patients – is not unrealistic. This is not like the most successful builder in the land now saying they wish to build castles in the air. There are many brakes in the system, other checkpoints that can be tampered with, other ways to unleash immune cells to fight cancer more effectively. Allison's work led to the first medicine of its kind, but it's not the only one.

In 1992, a different protein receptor on T cells was discovered by Japanese scientist Tasuku Honjo. Proteins are encoded by genes, and this particular protein was brought to Honjo's attention as a result of a search for genes that cause cells to die. It was therefore given another cumbersome name – programmed cell death 1 or PD-1.[57] In fact, it had been misnamed – it eventually turned out that this receptor has nothing to do with T cells dying – and for many years, the role of the PD-1 receptor remained mysterious. A big clue came when mice were genetically altered to lack the gene which encodes for the PD-1 receptor. Without PD-1, the mouse immune system reacted more vigorously; immune cells multiplied more when stimulated, and some of the mice, especially those that were elderly, spontaneously developed an autoimmune disease.[58] This fitted with the idea that PD-1 also sends a switch-off signal to immune cells – another brake on the system – and that without it, the immune system is more reactive, overly so when autoimmune disease develops.

We now know that after being switched on to participate in an immune response, all kinds of immune cells, including T cells, present the PD-1 receptor protein at their surface.[59] This receptor locks onto proteins on the surface of other cells that have been exposed to cytokines released as part of the immune response. Once its PD-1 receptor protein is engaged in this way, a switch-off signal is triggered and the immune cell ceases its response. In this way, PD-1 is instrumental in stopping a reaction from being overly aggressive or going on for too long.

To that extent, the roles of PD-1 and CTLA-4 overlap – both serve as brakes on the immune response – but they act in different circumstances. The proteins which PD-1 lock onto are induced in cells exposed to inflammation while CTLA-4 locks onto proteins on other immune cells, such as dendritic cells. An implication of this is that PD-1 may be particularly important in dialling down an ongoing local immune response, while CTLA-4 is perhaps more important in dampening the system as a whole in order to prevent a body-wide type of autoimmune disease. Understanding the complementary roles of different immune system brakes is at the frontier of research but from what we already know, blocking PD-1 might be especially potent in boosting a local anti-tumour response, unleashing immune cells that have managed to infiltrate a tumour but which are being held in check there by the PD-1 brake.

The programme to develop a medicine that blocked PD-1 benefited from what had been learnt by blocking CTLA-4.[60] It had already been established, for example, that it may take some time for patients to clearly benefit, and, perhaps most importantly, the success of blocking CTLA-4 meant that all major pharmaceutical companies around the world were keen to be involved. Clinical trials soon established that blocking PD-1 was even more effective than blocking CTLA-4 in melanoma patients, and led to fewer adverse side effects.[61] Other difficult-to-treat cancers were shown to succumb to attack by the immune system when the PD-1 brake was taken off.[62] Scientifically, this meant that the successes derived from blocking CTLA-4 were no mere fluke. The idea behind it – that we can fight disease by making sure the immune system doesn't turn itself off – was right.

This is still only a beginning. We now know of over twenty other brake receptors in the immune system.[63] Most of these switch off specific types of immune cell: Natural Killer cells, macrophages, dendritic cells, T cells, B cells or others. We must now test, in academic labs and companies large and small, whether or not antibodies that block these receptors, individually or in combination, will unleash immune cells to tackle different types of cancer. And not only cancer: immune cells also switch off after fighting long-term viral infections, such as HIV. So taking the brakes off may help unleash the immune system against long-term infectious diseases.

Unfortunately, we can't predict which types of cancer or other diseases will be most affected by taking the brakes off a particular type of immune cell; the system is too complex and our understanding too slight. But we have identified many brakes in the system and have the technology to shut them off one by one. 'We don't know everything,' says Eric Vivier, who co-founded a company seeking new checkpoint inhibitors, 'but maybe we know enough … Everything you have to do [next] is a bet.'[64] His company, born out of friendships that developed between a few of the scientists who identified and studied brake receptors on Natural Killer cells, has placed a large bet on blocking receptors on these cells in particular. Whether or not their bet will pay off is impossible to know, but for humankind as a whole, having multitudes of companies blocking all kinds of receptors on various immune cells, either alone or in combination, and testing what happens in different diseases – in other words, placing lots of bets – seems a sensible strategy.

At the level of patients, however, as new checkpoint inhibitors are discovered, it becomes ever more important to find ways of determining in advance who has the best chance of responding to them. Simply trying each one in turn is too crude. In order to avoid administering a checkpoint inhibitor to a patient who is potentially predisposed to serious side effects from it and to ensure instead that they are given the one most likely to work, we need ways to find out precisely what is happening inside the patient beforehand. The various measures used for this purpose are referred to in the jargon as 'biomarkers'. The one familiar biomarker, often checked in hospitals, is a blood count; a glance at the number of cells in a drop of blood might reveal if someone has anaemia or an infection. But a simple blood count is broad-brush and imprecise. In the case of checkpoint inhibitors, we need biomarkers that are far more precise.

One potential biomarker would be to test which brakes have been switched on in a patient by testing which brake receptors are present at the surface of a person's immune cells. This would allow us to select a checkpoint inhibitor that targets those particular receptors. A person's tumour can also be analysed to determine whether or not it contains the protein molecules that trigger particular brake receptors on immune cells. This could, in principle, predict whether or not blocking the PD-1 brake system, for example, is likely to benefit a patient. Unfortunately, this has not proved to be so easy, and predicting patient responses this way has become controversial.[65] First, the brakes are dynamic; knowing what is keeping the immune system in check one day might not reflect the situation the next day. Also, things can change as a result of treatment; as one brake comes off thanks to a checkpoint inhibitor, a tumour may adapt to exploit another brake system. As well as this, immune cells and cancer cells are highly variable, even within a single patient. A single tumour is sometimes said to be not a single disease but over a million different ones, with each of its millions of cancer cells being slightly different. The quest for predictive biomarkers is important but as a research field it is still young.

The search for biomarkers may in fact soon lead us to other problems. While few people would argue against knowing in advance whether or not blocking the PD-1 brake system can help a cancer patient, it's not a great leap from this use of biomarkers to using them in order to characterise a person's immune system long before a problem has become apparent. What if we could check the state of a person's immune system in detail – to test, for example, whether or not an individual has specific immune cells that are likely to cause autoimmune disease in old age; to monitor changes in, say, the number of regulatory cells a person has? As well as predicting who would be most likely to benefit from a particular medicine, this might give us an accurate assessment of someone's overall health and allow us to predict which illnesses a person is especially susceptible to. Genetic research and genetic testing especially have caused a great deal of debate for fear that they will lead to the social engineering of Huxley's *Brave New World*. Yet we may still end up there through an unexpected back door: the science of the immune system.

One reason this might not happen, irrespective of any governmental legislation, is that the system could just be too complex to be predictable. A closer look at how the immune system's brakes actually work shows just how complex the system is. One way that CTLA-4 puts a brake on T cells is by locking onto and covering up stimulatory proteins on other cells – effectively masking the immune system's alarm signals. But as well as this, CTLA-4 is able to grab co-stimulatory proteins on other immune cells and not just mask them but rip them off and have them destroyed, effectively hoovering up the system's alarm signals altogether.[66] And that's not all. In fact, as well as being a brake on immune cells, CTLA-4 is also an accelerator, in the sense that it is able to make immune cells move faster.[67] This has at least two consequences. First, it reduces the contact time between immune cells, which reduces their ability to interact and dampens the overall response.[68] Second, it makes it harder for fast-moving immune cells to hold onto a cancer cell long enough to kill it.[69] Presumably, blocking CTLA-4 helps cancer patients by stopping some of these processes. But it perhaps does something else too.

Allison used an antibody to block the CTLA-4 receptor, to stop it from working. However, antibodies that are naturally produced in the human body do more than merely block whatever it is that they adhere to. Antibodies are Y-shaped protein molecules which, as part of our natural immune defence, lock onto germs or diseased cells. The double-pronged end of the Y sticks onto a germ or diseased cell, while the back end is left exposed. Immune cells have receptors that fit the back end of an antibody and when this happens, the immune cell is triggered to kill or engulf whatever it is that the front end is tethered to. This means that while the front end of Allison's therapeutic antibody might block the CTLA-4 receptor from working, the back end of the antibody could, at least in principle, attract immune cells to simply destroy the T cells that the antibody has stuck to. Killing off the body's immune cells doesn't, on the face of it, seem like a useful thing to do in order to boost an anti-tumour response. But there's an important twist here.

Regulatory T cells – the guardian cells that Shimon Sakaguchi discovered, discussed in Chapter Seven – have high levels of CTLA-4 at their surface. So Allison's antibody could, in theory, lock onto regulatory T cells and tag them for destruction. Whether or not this is what happens in patients is controversial. But if this is correct then Allison's antibody might lift the brakes off the immune system in a way that is entirely different to how he thought it would: it may work in part by triggering the destruction of regulatory T cells.

Overall, the precise way in which an antibody against CTLA-4 helps cancer patients is not entirely clear; it may well have many effects.[70] At one level this doesn't matter; what matters are the worlds saved. But at another level, how it works *is* important – not as a matter of academic interest, not for the satisfaction of curiosity, but because understanding more deeply how the immune system's brakes work and how these therapeutic antibodies help patients might allow us to tweak their design, boost their efficacy, know which patients to treat, and create alternatives that target other molecules involved in the same process.

One way to realise Allison's wish – to boost the success of checkpoint inhibitors – is to use them not just on their own, but alongside other medicines. A combination of four – a tumour-targeting antibody, a cytokine, a vaccine and a check-point inhibitor – has been shown, in mice, to eradicate large established tumours which were otherwise untreatable.[71] Each medicine has a modest effect alone but together they became a cure. Combinations of different medicines are almost certain to be useful in cancer patients but the problem is in finding the right mix. There are an enormous number of permutations that could be tried and each component will have its own dosing and timing requirements, which may also vary depending on which other medicines are included in any given regimen. One approach is to place lots of bets, but there aren't even enough patients to try a fraction of all the possible combinations of drugs out there. We can't rely on chance; we also need a strategy, which requires different academic institutes to work together.

Enter Sean Parker, the billionaire entrepreneur, co-founder of the music-sharing service Napster and the first president of Facebook, played by Justin Timberlake in the movie *The Social Network*. In 2016, at age thirty-six, he donated $250 million to set up the Parker Institute, a collaboration of over forty labs across six US cancer centres or, in his words, 'a Manhattan Project for curing cancer with the immune system'.[72] He launched the institute with a party at his multimillion-dollar home in Bel Air, Los Angeles. Guests included actors Tom Hanks, Goldie Hawn and Sean Penn, magician David Blaine, comedian James Corden and musicians the Red Hot Chili Peppers, Katy Perry and Lady Gaga. The razzmatazz brought cancer immunotherapy to all corners of the media. 'Who doesn't like a huge infusion of cash? It's a wonderful thing', Tom Hanks said to one TV crew. 'Sean Parker's taking six organisations and making them work as one so that they will be sharing all the information about their cancer research and immunotherapy down the line, and it's going to alter the way cancer is studied and the way people with cancer are treated. Pretty good thing don't you think?'[73]

Parker, reportedly worth $3 billion,[74] was driven to fight cancer in part because one of his close friends, Laura Ziskin, a movie producer whose work included *Pretty Woman* and *The*

Amazing Spider-Man, died from cancer in 2011, aged sixty-one. Parker says he wants to hack cancer just like he hacked music. To him, a hack is 'a clever work-around or a clever way of leveraging an existing system to do something you didn't think it was going to be able to do' – which for him, makes immunotherapy a hack.[75] Cancer is more complicated than the music industry but Parker's attitude is welcome. 'Hackers share certain values,' he wrote in the *Wall Street Journal*, 'an anti-establishment bias, a belief in radical transparency, a nose for sniffing out vulnerabilities in systems, a desire to "hack" complex problems using elegant technological and social solutions, and an almost religious belief in the power of data to aid in solving those problems.'[76]

As well as providing money, the big change that the Parker Institute aims to make is to alter the way cancer research is organised. Traditionally, researchers across different cancer centres would have to compete with each other for government funding. This meant that there was an incentive to keep data secret for the many months it takes for applications to be assessed. With communal funding in place from Parker, sharing ideas is immediately made easier. Intellectual property generated in any of the six centres involved is also shared and an overarching body helps ensure that discoveries are not then lost to small start-up companies that collapse, or shelved by larger pharmaceutical companies whose priorities can quickly shift. Profits generated from intellectual property are to be shared among the individual discoverers and all the cancer centres involved. An individual centre could feasibly get more money by keeping their own intellectual property, but Parker persuaded the different cancer centres to sign up to sharing because it gives everyone a greater chance of winning something.[77]

Once rivals, Bluestone and Allison now work together under the umbrella of the Parker Institute. Perhaps it helps that all the institute leaders are in the second half of their lives, already hugely successful. Retreats encourage brainstorming and the atmosphere is borrowed from Silicon Valley: take risks, fail early, fail forward.[78] The paperwork has also been streamlined. For any company wanting to test a new medicine they only need to sign a single agreement with the Parker Institute to begin work across all six cancer centres. Previously, there would be many months of delay while a contract was negotiated with each centre individually.[79]

Bluestone is the institute's first overarching president. He says that Parker has provided the resources needed to try things differently, to create a sandbox where people trust each other and share ideas. 'Frankly,' he says, 'we're in an amazing revolution, [a] biomedical revolution. It's not unlike the revolution of the late 1800s to early 1900s in the Industrial Revolution. We have more access to knowledge, we have more access to tools, so you have more access to great ideas than ever before in biomedical science … The technology of the immune cell is enormously complex but we're starting to be able to decode that, and we're starting to be able to utilise it to attack cancer.'[80]

This revolution he talks of involves more than just checkpoint inhibitors; there are now hundreds of branches of immunotherapy. One of the areas where the Parker Institute aims to make a difference is in testing whether or not different ideas can be combined. One reason that checkpoint inhibitors do not work for everyone is probably because they work best at unleashing an immune response that is already present in a person – which implies that they would not work well for cancers which have relatively few mutations and are less visible to the immune system. One way of tackling this problem is to combine checkpoint inhibitors with another treatment that makes absolutely certain that a patient has immune cells capable of detecting their cancer.

So how might this be done? Recall from Chapter Three how Steven Rosenberg attempted to treat cancer in the 1980s.[81] He isolated immune cells from patients, boosted their activity in a lab dish (with cytokines) and then infused them back into patients. This was occasionally successful but there were serious problems with side effects. One reason the method did not work especially well is likely to be that the batch of immune cells grown in a lab dish contained immune cells of many different types, only a small number of which would have been able to attack the tumour. In 2011, Carl June at the University of Pennsylvania used a more sophisticated approach – and cured a patient of their leukaemia.[82]

Like Rosenberg before him, he extracted T cells from the patient, but before being infused back into the patient the T cells were genetically manipulated in such a way that a new receptor was added to them, one which targets the patient's cancer. This is called CAR T cell therapy, named for the added receptor – CAR stands for chimeric antigen receptor – which is made up of the front end of an antibody that locks onto cancer cells and a back end that triggers a T cell to kill. In this way, a patient's T cells are reprogrammed to specifically target and kill their own cancer cells. In fact, the idea for this innovation had been suggested in 1989, over two decades before it finally became a medical success.[83] One reason it took this long was the time needed to develop a successful procedure for inserting an additional receptor into T cells. Ultimately, June used a disabled version of the HIV virus to do so, harnessing the virus's natural ability to infect T cells and insert a copy of its genes in any cell it infects.[84]

June and his colleagues had hoped that CAR T cell therapy might bring benefit to cancer patients but did not dare dream of a complete remission. Two of the first three patients achieved such remission. One, a sixty-five-year-old scientist with chronic lymphocytic leukaemia, was treated with a dose of 14 million genetically modified T cells. He wrote anonymously for the University of Pennsylvania's web page: 'I am still trying to grasp the enormity of what I am a part of – and of what the results will mean to countless others with Chronic Lymphocytic Leukemia or other forms of cancer. When I was a young scientist, like many I'm sure, I dreamed that I might make a discovery that would make a difference to mankind – I never imagined I would be part of the experiment.'[85] On 30 August 2017, the FDA approved the use of CAR T cells for one type of cancer and a few weeks later for another, making this living drug unquestionably revolutionary.

The optimal version of this type of therapy is what is needed now. There are many parameters to work through: we don't yet know which molecules on a person's cancer are the best to target, we don't yet know whether or not every cancer cell would have to possess the same signature, we don't yet know how to limit the possibility of healthy bystander cells being attacked, causing unwanted side effects, and so on. At least in principle, this type of therapy could be used more widely than in cancer patients. For example, CAR T cells can also be engineered to kill off the fraction of a person's immune cells that are causing an autoimmune disease.[86] One of the greatest problems is the toxicity of genetically enhanced immune cells. June's version of T cell therapy was more sophisticated than Rosenberg's, but one day June's work will, in all likelihood, also seem relatively crude.

June, Bluestone and others at the Parker Institute are seeking to combine the idea of CAR T cell therapy with the idea of checkpoint inhibitors. New technology for genetically manipulating cells makes it easy for the same immune cells to be tweaked in more than one way.[87] The new plan is to engineer the same T cells not only to have a receptor that can recognise a person's cancer but also to lack a brake system. That way, it is hoped, the enhanced T cells will fight for longer in the body. And because these T cells will be engineered to recognise the person's cancer cells directly, it is hoped that taking the brakes off these cells will cause fewer side effects compared with treatments that unleash the immune system more broadly.

Other new cancer treatments have emerged from unlikely places. New versions of thalidomide, for example – the shadowy drug once used to alleviate morning sickness but responsible for thousands of babies dying or being born with malformed limbs – can also enhance the immune system's ability to attack cancer cells. Thalidomide was discovered in early 1954 as a by-product in a search for a cheaper way to manufacture antibiotics. The German pharmaceutical company Grünenthal tested the new chemical in many ways, seeking a disease to combat with it, which eventually led to its use as a sedative and to prevent morning sickness. Over 10,000 infants were deformed by thalidomide, mutilated by what was supposed to be a medicine.[88] In 1962, the drug was banned worldwide. A few years later, however, anecdotes emerged that thalidomide could help patients with a particular complication of leprosy.[89] Studying this, it became clear that thalidomide has many effects on the human body, including on the immune system. US company Celgene created a safer derivative of thalidomide, now sold as Revlimid, which can help treat at least one type of cancer, multiple myeloma.[90] This medicine has kept my own father, afflicted with myeloma, alive for many years. Research in my lab has helped understand how this derivative of thalidomide works, helping to seed ideas for even better versions. It operates in many ways, but one thing it does is lower the threshold at which the body's Natural Killer cells can be switched on to attack cancer cells.[91]

We are at the dawn of a health revolution. But the elephant in the room is global poverty. Nearly half the world lives on less than $2 a day, and the economics of making and providing new medicines is another kind of tragedy. Our lacklustre effort to develop a vaccine for the Ebola virus since its discovery in 1976 is a case in point; the virus was hardly studied until richer countries felt threatened.[92] Though it deals in matters of life and death, the pharmaceutical industry is a business, not a charity, and when deciding research priorities, the potential for financial gain is a critical factor, if not the decisive factor. Although this is a book about ideas, science, history and our trajectory as a species, it would be deceptive – dishonest – to write about new medicines without mentioning the financial problems that stand in our way: we sorely need new international institutes and different ways of funding medical research and medicines, where the well-being of humanity, and other life on earth, is paramount and financial profit irrelevant. I hope this is the brave new world that awaits us.

As with all scientific revolutions, new knowledge isn't all that counts. We will be judged by our children and grandchildren, not by what we know, but by what we do with what we know.

Epilogue

Science is many things. A method, a journey, a route to power, a body of knowledge, a thing you loved or hated at school, a jigsaw puzzle with an infinite number of pieces, a force for good and bad which has produced both food and bombs. Arguably, its greatest success has been, and will be for some time to come, in curing diseases.

Still, our body's own remedy – the immune system – is far more powerful than any medicine we have devised. Most germs are tackled by our body without us hardly knowing. Over many decades, we have begun to unpick how this works – by testing what happens when one type of cell is missing or is in abundance, when a gene is inactivated or enhanced, when a chemical pathway is switched on or off. Step by step, with an occasional misstep, we have uncovered many of our immune system's secrets. But like the solar system, the financial system – any big system – the immune system remains enigmatic. All the theories have cracks in them, each idea works only in certain situations, everything is only some kind of an approximation. I and thousands of others continue to devote our days and lives to solving the puzzles that remain. Someday we may find a grand unified theory of the immune system, a few principles that capture precisely how it all works, or one diagram that fits on a T-shirt. But that dream may never work out. And it might even be the wrong thing to aim for.

Perhaps the immune system – and human biology in general – is more pluralistic than we currently tend to imagine. Physicists have long since learnt to live with the strange fact that light sometimes behaves like a wave but at other times as a particle. How we think of light varies according to the measurement being made; any one view is too narrow. It seems to me that the complexity of the immune system means that it also acts *strangely*. Its behaviour can't be explained in a simple-to-say sound bite, such as that the immune system is set up to look for danger or to discriminate self from non-self. These are useful ideas, stronger than metaphors, but weaker than laws that govern every situation. Since the immune system, like all of human biology, evolved without any principled foundation, it might be futile to seek one.

In the near future, irrespective of whether or not we can describe the system in simple terms, we may be able to make predictions about one's health by taking a few precise measure-ments – a detailed analysis of different blood cells, say – and plugging that information into a mathematical or computational simulation (in the same way that mathematics provides the only way to describe how light behaves). In the meantime, using the knowledge we have now, we can make smart lifestyle choices and create new medicines to fight infections, cancer, auto-immune diseases and other such maladies. But death will not be eradicated and paradise is not promised. From here, we must step forward carefully, thinking deeply about what we set out to treat. Not all deaf people want to join the hearing world. And how shocking it feels now to know that in the UK in the 1950s and 60s, homosexuality was seen as a condition to be treated with the female hormone oestrogen or electric shocks. Science is many things but there are some things it must not be: it must not be the pursuit of any supposed perfection in the human body.

For me, and I suspect many others, there's something else we gain from fathoming the details. We see ourselves for what we truly are, which lifts us out of the humdrum. A crucial point of science – the knowledge and the journey – is that it's our escape by ascent.

Acknowledgements

I am especially grateful to those I had the privilege of interviewing for this book, including: Arne Akbar, Bruce Beutler, Jeffrey Bluestone, Leslie Brent, Brian Crucian, Kathryn Else, Marc Feldmann, Jordan Gutterman, Jules Hoffmann, Matthew Krummel, Lewis Lanier, Bruno Lemaitre, Janet Lord, Andrew Loudon, Andrew MacDonald, Ravinder Maini, Ofer Mandelboim, Polly Matzinger, Ruslan Medzhitov, Werner Müller, Christian Münz, Luke O'Neill, Peter Openshaw, Fiona Powrie, David Ray, Akhilesh Reddy, Shimon Sakaguchi, Mark Travis, Jan Vilček, Eric Vivier and Santiago Zelenay.

I am also indebted to many others who helped me address specific issues while writing this book, including: Walter Bodmer, Thiago Carvalho, Matthew Cobb, Alasdair Coles, Francesco Colucci, Stephen Eyre, Leroy Hood, Jonathan Howard, Tracy Hussell, John Inglis, Pippa Kennedy, Philippa Marrack, Steve Marsh, David Morgan, Vern Paetkau, Eleanor Riley, Christopher Rudd, David Sansom, Matthew Scharff, Kendall Smith, Robert Sinden and Jan Witkowski. I am hugely grateful to those who commented on early versions of some or all the text, including: Veronica Bartles, Doreen Cantrell, George Cohen, Mark Coles, Siamon Gordon, Salim Khakoo, Andrew MacDonald and Ofer Mandelboim. Importantly, any remaining errors in this book are my responsibility alone.

I thank my parents, Marilyn and Gerald Davis, for their enduring support. I also thank Matthew Cobb, Armand Leroi and Peter Parham for their encouragement, and members of my research team who have guided my thinking over many years. I'm also indebted to the leadership at the University of Manchester, including Nancy Rothwell, Martin Humphries, Ian Greer and Tracy Hussell, as well as others, earlier in my career at Imperial College London, including Maggie Dallman, Murray Selkirk and Ian Owens, for fostering an environment that allows me to study and write about the immune system. I also thank Jack Strominger, whose lab at Harvard University was where I began to study the immune system.

My editor, Will Hammond at Bodley Head, has had a vital and wonderful influence on this book, its overall shape as well as the final text. I'm also grateful to David Milner who copy-edited the text with great skill. Caroline Hardman, my literary agent at Hardman and Swainson, has been enormously helpful right from the very beginning. I am also hugely grateful for the input of my editor at Doubleday Canada, Tim Rostron. Above all, I thank my wife Katie, and our children Briony and Jack, for sharing the journey.

Notes

Overview

1. This interview features in the BBC TV programme *The Pleasure of Finding Things Out*, part of the Horizon series, archived online here: http://www.bbc.co.uk/iplayer/episode/p018dvyg/horizon-19811982-9-the-pleasure-of-finding-things-out. A transcript of the interview is also available in Jeffrey Robbins (ed.), *The Pleasure of Finding Things Out: The Best Short Works of Richard P. Feynman* (Penguin, 2001) – but from a transcript alone, you miss Feynman's absorbing oratory skill. Feynman's story is captured wonderfully in James Gleick, *Genius: Richard Feynman and Modern Physics* (Abacus, 1992).

2. Irwin, M. R., 'Why sleep is important for health: a psychoneuroimmunology perspective', *Annual Review of Psychology* **66**, 143–72 (2015).

3. Dorshkind, K., Montecino-Rodriguez, E., & Signer, R. A., 'The ageing immune system: is it ever too old to become young again?', *Nature Reviews Immunology* **9**, 57–62 (2009).

Chapter One

1. Bilalić, M., McLeod, P., & Gobet, F., 'Inflexibility of experts – reality or myth? Quantifying the Einstellung effect in chess masters', *Cognitive Psychology* **56**, 73–102 (2008).

2. There are many experiments that show the 'Einstellung' effect – a significant field of research in its own right. A great entry point to this subject is Bilalić, M., & McLeod, P., 'Why good thoughts block better ones', *Scientific American*, **310**, 74–9, March 2014.

3. Matzinger, P., 'Charles Janeway, Jr, Obituary', *Journal of Clinical Investigation* **112**, 2 (2003).

4. Gayed, P. M., 'Toward a modern synthesis of immunity: Charles A. Janeway Jr and the immunologist's dirty little secret', *Yale Journal of Biology and Medicine* **84**, 131–8 (2011).

5. Janeway, C. A., Jr, 'A trip through my life with an immunological theme', *Annual Review of Immunology* **20**, 1–28 (2002).

6. Ibid.

7. *State of the world's vaccines and immunization* (third edition, World Health Organization Press, 2009).

8. Variolation is a term often used to describe inoculation with smallpox. Variolation can be defined as an application of a small dose of infection in a controlled manner, whereas vaccination can be distinguished from this as the use of dead or weakened infection. The terms inoculation, vaccination and immunization may also be given subtly different meanings. However, the various ways in which modern vaccines work makes, to my mind, specific definitions difficult for these terms and so I have used these terms interchangeably here.

9. Rhodes, J., *The End of Plagues: The Global Battle against Infectious Disease* (Palgrave Macmillan, 2013); De Gregorio, E., & Rappuoli, R., 'From empiricism to rational design: a personal perspective of the evolution of vaccine development', *Nature Reviews Immunology* **14**, 505–514 (2014).

10. Silverstein, A. M., *A History of Immunology* (second edition, Academic Press, 2009).

11. A very brief history of the Royal Society is available online here: http://royalsociety.org/about-us/history/

12. Mead, R., *A Discourse on the Small Pox and Measles* (John Brindley, 1748). This book is by Richard Mead, the prominent London physician who carried out the inoculations of prisoners in 1721. The story of the royal experiment is told in Chapter Five, 'Of the inoculation of the small pox'.

13. Before inoculating their own children, the Princess of Wales first paid to have five orphan children inoculated. The test on prisoners had involved only adults and she felt it important to test the safety of the procedure in other children, before her own.

14. It's worth bearing in mind that celebrities influence public opinion quite independently of the prevailing orthodox view

of, for example, relevant scientific bodies. One example is the story of Jenny McCarthy, former *Playboy* model and partner of actor Jim Carrey, who has claimed that her own son, Evan, developed autism as a result of vaccination. Her opinions reached a huge audience around 2007–9, for example by her appearing on Oprah Winfrey's TV show. Her personal story is emotive and she has said: 'my science is Evan. He's at home. That's my science.' She has written several books herself, including *Louder than Words: A Mother's Journey in Healing Autism* (Plume, 2008). Her story is also discussed in Mark A. Largent, *Vaccine: The Debate in Modern America* (Johns Hopkins University Press, 2012), 138–48.

15. Silverstein.
16. Jenner, E., *An Inquiry Into the Causes and Effects of the Variolae Vaccinae: A disease discovered in some of the Western Counties of England, particularly Gloucestershire, and known by the name of the cow pox* (1798). This landmark text has been republished many times and the full text is freely available on the web, such as here: http://www.bartleby.com/38/4/1.html.
17. The word vaccine was coined by surgeon Richard Dunning. The use of the word 'vaccine' in situations other than for the use of cowpox to protect against smallpox has been attributed to Louis Pasteur (1822–95).
18. The fact that smallpox was tackled on a global scale is important. 'Probably the worst thing that ever happened to malaria in poor nations,' journalist Tina Rosenberg has written, 'was its eradication in rich ones', quoted on p. 44 of Eula Biss's wonderful book about vaccination, *On Immunity* (Graywolf Press, 2014).
19. Rhodes.
20. Although not known in 1920s, it was well established in 1989 that the immune system learns what our body is made of at an early age, so that it is then ready to attack anything else. This is discussed in much greater depth in my first book, *The Compatibility Gene* (Allen Lane, 2013).
21. Oakley, C. L., 'Alexander Thomas Glenny. 1882–1965', *Biographical Memoirs of Fellows of the Royal Society* 12, 162–80 (1966).
22. Oakley, C. L., 'A. T. Glenny', *Nature* 211, 1130 (1966).
23. Ibid.

24. Having been brought up in an especially conservative Christian household in which going to the theatre or concerts was forbidden, Glenny had little interest in anything outside work.

25. Marrack, P., McKee, A. S., & Munks, M. W., 'Towards an understanding of the adjuvant action of aluminium', *Nature Reviews Immunology* **9**, 287–93 (2009).

26. Gura, T., 'The Toll Road', *Yale Medicine* **36**, 28–36 (2002).

27. The Cold Spring Harbor Symposia on Quantitative Biology is a series of conferences which began in 1933. Many renowned scientists were present at the 1989 meeting, including Tasuku Honjo, Leroy Hood, John Inglis, Richard Klausner, Fritz Melchers, Gustav Nossal and Rolf Zinkernagel. Twenty photos from the meeting have been archived and are available online here: http://libgallery. cshl.cdu/items/browse/tag/Immunological+Recognition. John Inglis, executive director of Cold Spring Harbor Press, has told me (email correspondence, 25 March 2015) that in his recollection, Janeway sent him the paper for inclusion in the meeting proceedings after the event and that at the meeting itself, he didn't give a formal talk on the subject. He may have just discussed the ideas with people informally at the meeting itself.

28. Janeway, C. A., Jr, 'Approaching the asymptote? Evolution and revolution in immunology', *Cold Spring Harbor Symposia on Quantitative Biology* **54 Pt 1**, 1–13 (1989).

29. Ibid.

30. Some scientists do not like the term 'pattern recognition', as this type of interaction between a protein and another molecule is usually called 'molecular recognition'. Nevertheless, the term 'pattern-recognition receptors' remains in common use today.

31. Charles Janeway, with colleague Paul Travers, first published their textbook *Immunobiology* in 1994. This and all subsequent editions have been hugely successful. The ninth edition of this book – now titled *Janeway's Immunobiology* – was published in 2016, updated by Kenneth Murphy and Casey Weaver.

32. George Bernard Shaw said this in a speech at a public dinner in London on 28 October 1930, to honour Albert Einstein. Excerpts from this speech are given in Michael Holroyd, 'Albert Einstein, Universe Maker', *New York Times*, 14 March 1991.

33. Calculated as $3 \times 100^4 = 3 \times 10^8$.

34. Calculated as seventy-two divisions in twenty-four hours (one every twenty minutes), leading to 2^{72} offspring.

35. Essentially, this means that the process of evolution by natural selection happens much faster for viruses than for us. For some viruses, this is enhanced even further because the rate at which genetic variations occur when viruses reproduce is far greater than it is in us (because the machinery used to copy genetic material is less careful in some viruses). This works for viruses because any defective offspring that may arise have little consequence overall.

36. Janeway (1989).

37. Email correspondence with Leroy Hood (10 February 2015) and Jonathan Howard (12 February 2015), respectively.

38. Janeway (1989).

39. Medzhitov, R., 'Pattern recognition theory and the launch of modern innate immunity', *The Journal of Immunology* 191, 4473–4 (2013).

40. Interview with Ruslan Medzhitov, 31 March 2015.

41. Ibid.

42. Ibid.

43. Gura.

44. Interview with Ruslan Medzhitov, 31 March 2015.

45. Ibid.

46. Dahl, R., *The Minpins* (Puffin, 1991).

47. McKie, R., 'Six Nobel prizes – what's the fascination with the fruit fly?', *Observer*, 8 October 2017. Available online here: https://www.theguardian.com/science/2017/oct/07/fruit-fly-fascination-nobel-prizes-genetics.

48. Interview with Jules Hoffmann, 7 April 2015.

49. Particularly inspirational to Hoffmann – and others – were Hans Boman's discoveries through the 1970s and early 1980s which culminated in identifying antibacterial peptides in North America's largest native silk moth, *Hyalophora cecropia*. Over 700 different antimicrobial peptides have since been isolated from mammals, as discussed by Jack L. Strominger in *The Journal of Immunology* 182, 6633–4 (2009). Hans Boman died on 3 December 2008.

50. Fehlbaum, P., et al., 'Insect immunity. Septic injury of Drosophila induces the synthesis of a potent antifungal peptide with sequence

homology to plant antifungal peptides', *Journal of Biological Chemistry* **269**, 33159–63 (1994).

51. O'Neill, L. A., Golenbock, D., & Bowie, A. G., 'The history of toll-like receptors – redefining innate immunity', *Nature Reviews Immunology* **13**, 453–60 (2013). This scholarly and authorative article reviews, in depth, the sequence of events that led to the discovery of toll-like receptors.

52. Lemaitre, B., 'The road to toll', *Nature Reviews Immunology* **4**, 521–7 (2004).

53. Lemaitre, B., Nicolas, E., Michaut, L., Reichhart, J. M., & Hoffmann, J. A., 'The dorsoventral regulatory gene cassette spatzle/Toll/cactus controls the potent antifungal response in Drosophila adults', *Cell* **86**, 973–83 (1996).

54. The three reviews of this paper, obtained by the journal *Cell*, are archived online by the first author Bruno Lemaitre. It is interesting that all three reviews were very encouraging although, as is nearly always the case with peer review, each scientist did demand several more experiments before they felt the work should be published. The reviews are available here: http://www.behinddiscoveries.com/toll/resources

55. Interview with Jules Hoffmann, 7 April 2015.

56. Ibid.

57. It's certainly plausible that Medzhitov did have the data in hand already; there were clues as to what to look for from other immune receptor pathways (involving IL-1 and TNF). When seeking to clarify some of these details, one impartial immunologist replied to me that 'it's all a bit cloak and dagger, Dan'.

58. Medzhitov, R., Preston-Hurlburt, P., & Janeway, C. A., Jr, 'A human homologue of the drosophila toll protein signals activation of adaptive immunity', *Nature* **388**, 394–7 (1997).

59. It's important to note, however, that Barbara Baker's work on the tobacco N gene, part of the plant's innate defence, began before the work on flies discussed here. Baker's research indicates a similarity in innate immunity defences even between mammals and plants.

60. Interview with Bruce Beutler, 21 April 2015.

61. A short autobiography by Beutler is available online at http://www.nobelprize.org/nobel_prizes/medicine/laureates/2011/beutler-bio.html.

62. Interview with Bruce Beutler, 21 April 2015.

63. Ibid.

64. Ibid.

65. Beutler's grandparents all moved to the USA to avoid persecution of Jews in Europe. Anti-Semitism shaped the ethos of his family and Beutler says, in autobiographical notes archived by the Nobel Committee, that 'probably we all felt a need to excel partly because of these facts; to show that we were as good as the other children in our schools'.

66. Beutler Nobel autobiography.

67. Ibid.

68. Poltorak, A., et al., 'Defective LPS signaling in C3H/HeJ and C57BL/10ScCr mice: mutations in Tlr4 gene', *Science* **282**, 2085–8 (1998).

69. O'Neill (2013).

70. Qureshi, S. T., et al., 'Endotoxin-tolerant mice have mutations in toll-like receptor 4 (Tlr4)', *The Journal of Experimental Medicine* **189**, 615–25 (1999).

71. Hoshino, K., et al., 'Cutting edge: toll-like receptor 4 (TLR4)-deficient mice are hyporesponsive to lipopolysaccharide: evidence for TLR4 as the Lps gene product', *The Journal of Immunology* **162**, 3749–52 (1999).

72. Beutler recounts how he heard the news in a telephone conversation with Adam Smith, editorial director of Nobel Media, recorded on the same day as the prize was announced, 3 October 2011: http://www.nobelprize.org/mediaplayer/index.php?id=1632.

73. Allison, J. P., Benoist, C., & Chervonsky, A. V., 'Nobels: Toll pioneers deserve recognition', *Nature* **479**, 178 (2011).

74. Paul, W. E., & Germain, R. N., 'Obituary: Charles A. Janeway Jr (1943–2003)', *Nature* **423**, 237 (2003).

75. Paul, W. E., 'Endless fascination', *Annual Review of Immunology* **32**, 1–24 (2014).

76. Interview with Ruslan Medzhitov, 31 March 2015.

77. Interview with Jules Hoffmann, 7 April 2015.

78. Ezekowitz, A., et al., 'Lawrence's book review unfair to Hoffmann', *Current Biology* **22**, R482 (2012).

79. Lemaitre, B., *An Essay on Science and Narcissism: How Do High-Ego Personalities Drive Research in Life Sciences?* (Copy Media, 2016).

80. Cyranoski, D., 'Profile: Innate ability', *Nature* **450**, 475–7 (2007).

81. It's not clear why insects, for example, don't have or need a more complex immune system akin to ours. Body size and longevity are sometimes invoked in explanations, or that, as anatomy evolved to be more complex, a more complicated immune defence also developed.

82. This was suggested by Rolf Zinkernagel speaking at the 64th Lindau Nobel Laureate Meeting, 1 July 2014, and is recounted by Jules Hoffmann in an interview recorded at that meeting, available online here: http://www.dw.de/tomorrow-today-the-science-magazine-2014–07–07/e-17717966–9798.

83. Interview with Jules Hoffmann, 7 April 2015.

84. Rees, M., *Our Final Century* (William Heinemann, 2003).

85. Marrack et al.

86. De Gregorio & Rappuoli.

87. Beutler Nobel autobiography.

88. Interview with Luke O'Neill, 16 March 2016.

89. Interview with Ruslan Medzhitov, 31 March 2015.

Chapter Two

1. Koestler, A., *The Act of Creation* (Hutchinson, 1964).

2. Nussenzweig, M. C., & Mellman, I., 'Ralph Steinman (1943–2011)', *Nature* **478**, 460 (2011).

3. Steinman, R. M., 'Dendritic cells: understanding immunogenicity', *European Journal of Immunology* **37 Suppl 1**, S53–60 (2007).

4. Steinman, R. M., & Cohn, Z. A., 'The interaction of soluble horseradish peroxidase with mouse peritoneal macrophages in vitro', *The Journal of Cell Biology* **55**, 186–204 (1972).

5. Mosier, D. E., 'A requirement for two cell types for antibody formation in vitro', *Science* **158**, 1573–5 (1967). This is the first paper clearly showing the need for so-called accessory cells in an immune reaction. The observation was made by studying the reaction of mouse immune cells against red blood cells from sheep.

6. Jolles, S., 'Paul Langerhans', *Journal of Clinical Pathology* **55**, 243 (2002).

7. Steinman, R. M., & Cohn, Z. A., 'Identification of a novel cell type in peripheral lymphoid organs of mice. I. Morphology, quantitation, tissue distribution', *The Journal of Experimental Medicine* **137**, 1142–62 (1973).

8. Simons, D. J., & Chabris, C. F., 'Gorillas in our midst: sustained inattentional blindness for dynamic events', *Perception* **28**, 1059–74 (1999).

9. The Invisible Gorilla video is available online here: http://www.theinvisiblegorilla.com/gorilla_experiment.html

10. Drew, T., Vo, M. L., & Wolfe, J. M., 'The invisible gorilla strikes again: sustained inattentional blindness in expert observers', *Psychological Science* **24**, 1848–53 (2013).

11. Snyder, L. J., *Eye of the Beholder: Johannes Vermeer, Antoni van Leeuwenhoek, and the Reinvention of Seeing* (W. W. Norton, 2015).

12. Lindquist, R. L., et al., 'Visualizing dendritic cell networks in vivo', *Nature Immunology* **5**, 1243–50 (2004).

13. The origin of this quote is unclear. Albert Szent-Györgyi used the line in his 1957 book *Bioenergetics* but he used quotation marks, indicating he borrowed the line from elsewhere. He won the Nobel Prize in Physiology or Medicine in 1937.

14. Steinman, R. M., 'Endocytosis and the discovery of dendritic cells' in Moberg, C. L. (ed.), *Entering an Unseen World* (Rockefeller University Press, 2012).

15. Pollack, A., 'George Palade, Nobel Winner for Work Inspiring Modern Cell Biology, Dies at 95', *New York Times*, 9 October 2008.

16. Porter, K. R., Claude, A., & Fullam, E. F., 'A Study of Tissue Culture Cells by Electron Microscopy: Methods and Preliminary Observations', *The Journal of Experimental Medicine* **81**, 233–46 (1945).

17. Moberg, C. L., *Entering an Unseen World: A Founding Laboratory and Origins of Modern Cell Biology 1910–1974* (Rockefeller University Press, 2012).

18. Steinman, R. M., 'Dendritic cells: from the fabric of immunology', *Clinical and Investigative Medicine* **27**, 231–6 (2004).

19. Steinman's children were interviewed by Dan Woog for the article 'Remembering Ralph Steinman', published online 26 October 2011, in a blog, available here: http://06880danwoog.com/2011/10/26/remembering-ralph-steinman/.

20. Albert Claude had earlier established the basic process of using a centrifuge to separate parts of cells. Claude's first experiments with a centrifuge were carried out in 1937 and he obtained four crude fractions of cells, for example, in 1941. Claude shared the Nobel Prize in Physiology or Medicine, in 1974, with Palade and

de Duve. Swedish scientist Theodor Svedberg developed the scientific use of centrifugation before this, showing that the method could be used to separate out different protein molecules. In 1926, the Nobel Prize in Chemistry was awarded to Svedberg for this work.

21. De Duve, C., 'Exploring cells with a centrifuge' (Nobel Lecture, 1974). Available online here: http://www.nobelprize.org/nobel_prizes/medicine/laureates/1974/duve-lecture.pdf

22. Nussenzweig, M. C., 'Ralph Steinman and the discovery of dendritic cells' (Nobel Lecture, 2011). Available online here: http://www.nobelprize.org/nobel_prizes/medicine/laureates/2011/steinman_lecture.pdf

23. Gordon, S., 'Elie Metchnikoff: father of natural immunity', *European Journal of Immunology* **38**, 3257–64 (2008).

24. Metchnikoff, I., 'Nobel Lecture 1908' in *Nobel Lectures in Physiology or Medicine 1901–1921* (Elsevier, 1967).

25. Vikhanski, L., *Immunity: How Elie Metchnikoff Changed the Course of Modern Medicine* (Chicago Review Press, 2016).

26. Metchnikoff, O., *Life of Elie Metchnikoff* (translated from French) (Houghton Mifflin Company, 1921).

27. Metchnikoff (1967).

28. Vikhanski.

29. Ibid.

30. Ambrose, C. T., 'The Osler slide, a demonstration of phagocytosis from 1876 Reports of phagocytosis before Metchnikoff's 1880 paper', *Cellular Immunology* **240**, 1–4 (2006).

31. Paul, W. E., 'Bridging innate and adaptive immunity', *Cell* **147**, 1212–15 (2011).

32. Tirrell, M., Langreth, R., & Flinn, R., 'Nobel laureate treating own cancer dies before award announced', *Bloomberg Business* (4 October 2011). Available online here: http://www.bloomberg.com/news/articles/2011-10-03/nobel-laureate-ralph-steinman-dies-3-days-before-prize-announced.

33. Guy Raz interviews Adam Steinman, Ralph's son, broadcast on National Public Radio, 3 October 2011. Available online: http://www.npr.org/2011/10/03/141019170/son-of-nobel-winner-remembers-his-father.

34. Nussenzweig, M. C., & Steinman, R. M., 'Contribution of dendritic cells to stimulation of the murine syngeneic mixed

leukocyte reaction', *The Journal of Experimental Medicine* **151**, 1196–212 (1980); Nussenzweig, M. C., Steinman, R. M., Gutchinov, B., & Cohn, Z. A., 'Dendritic cells are accessory cells for the development of anti-trinitrophenyl cytotoxic T lymphocytes', *The Journal of Experimental Medicine* **152**, 1070–84 (1980).

35. Nussenzweig, M. C., Steinman, R. M., Witmer, M. D., & Gutchinov, B., 'A monoclonal antibody specific for mouse dendritic cells', *Proceedings of the National Academy of Sciences USA* **79**, 161–5 (1982).

36. Van Voorhis, W. C., et al., 'Relative efficacy of human mono-cytes and dendritic cells as accessory cells for T cell replication', *The Journal of Experimental Medicine* **158**, 174–91 (1983); Steinman, R. M., Gutchinov, B., Witmer, M. D., & Nussenzweig, M. C., 'Dendritic cells are the principal stimulators of the primary mixed leukocyte reaction in mice', *The Journal of Experimental Medicine* **157**, 613–27 (1983). In experiments reported in these papers in 1983, Steinman studied different kinds of immune reaction, including the reaction which occurs when blood cells from different people are mixed together. It's a reaction that can happen in a bone-marrow transplant situation if certain genes aren't well matched. The magnitude of the immune response, or how powerful the reaction is, can be monitored in several ways, such as testing how well the immune cells multiply to increase in number. Steinman's team showed that dendritic cells were at least 100–300 times more potent in triggering this kind of immune reaction, in comparison to any other immune cell.

37. Van Voorhis, W. C., Hair, L. S., Steinman, R. M., & Kaplan, G., 'Human dendritic cells. Enrichment and characterization from peripheral blood', *The Journal of Experimental Medicine* **155**, 1172–87 (1982).

38. Steinman (2004). Like many scientists who trained in Ralph Steinman's lab, Gerold Schuler went on to have his own stellar career in science. He become a head of department at Universitätsklinikum Erlangen, Germany, and has made many contributions in exploring the potential medical use of dendritic cells.

39. Schuler, G., & Steinman, R. M., 'Murine epidermal Langerhans cells mature into potent immunostimulatory dendritic cells in vitro', *The Journal of Experimental Medicine* **161**, 526–46 (1985).

40. Today, each dendritic-cell meeting attracts about 1,000 participants. The first meeting was held in Japan, 1990, as a satellite meeting. The second, held in the Netherlands in 1992, was a specific meeting dedicated to dendritic cells. This second meeting had 220 participants and fifteen invited speakers.

41. T cells which have a receptor which can easily trigger a reaction against healthy cells have been killed off (in the thymus) so that T cells in the lymph node do not react to the body's own components.

42. Often given less attention than, say, cancer or HIV, parasites affect well over a billion people and cause huge social and economic problems, arguably trapping some whole countries in poverty.

43. Anthony, R. M., Rutitzky, L. I., Urban, J. F., Jr, Stadecker, M. J., & Gause, W. C., 'Protective immune mechanisms in helminth infection', *Nature Reviews Immunology* 7, 975–9 (2007).

44. Kapsenberg, M. L., 'Dendritic-cell control of pathogen-driven T-cell polarization', *Nature Reviews Immunology* 3, 984–93 (2003).

45. Reis e Sousa, C., 'Dendritic cells in a mature age', *Nature Reviews Immunology* 6, 476–83 (2006). Essentially, I have described the basic model for how dendritic cells work in the body to trigger an immune reaction but many exceptions and details are discussed, for example, in this academic article.

46. Lamott, A., *Bird by Bird: Some Instructions on Writing and Life* (Pantheon Books, 1994).

47. Formally, there are class I and class II versions of the MHC proteins. Class I proteins are found on nearly all types of cells, but class II proteins are found only on some types of immune cells – known as antigen-presenting cells. The antigen-presenting cells include macrophages and dendritic cells and these are the types of cells able to trigger immune responses. Dendritic cells are the most potent antigen-presenting cells.

48. Davis, D. M., *The Compatibility Gene* (Allen Lane, 2013).

49. An additional signal that guides immune cell behaviour comes from soluble factors – cytokines – and this is sometimes referred to as Signal Three. Cytokines will be discussed in more detail in the next chapter.

50. The co-stimulation of T cells is a complex field in its own right. For a more detailed analysis of this topic, see Chen, L., & Flies,

D. B., 'Molecular mechanisms of T cell co-stimulation and co-inhibition', *Nature Reviews Immunology* 13, 227–42 (2013).

51. Like so much of immunology, there are exceptions to this statement. These same 'co-stimulatory proteins' can also engage inhibitory receptors on T cells to switch them off. The thinking here is that this serves to end the immune response after some time. In other words, the co-stimulatory proteins initially help switch on T cells, but in time they are also involved in switching T cells off when an immune response is no longer needed.

52. Ralph Steinman recalls the motivation behind his work on dendritic cells in an interview for the journal *Immunological Reviews*, recorded in March 2010. It is available online here: https://www.youtube.com/watch?v=BAn8wEpURtE.

53. Kool, M., et al., 'Cutting edge: alum adjuvant stimulates inflammatory dendritic cells through activation of the NALP3 inflammasome', *The Journal of Immunology* 181, 3755–9 (2008).

54. Interview with Kayo Inaba upon winning the 2014 L'Oréal-UNESCO Award for Asia-Pacific. Video available online here: https://youtube.com/watch?v=pd2tSDy8A3s.

55. In Japan, before working with Steinman, she had already studied dendritic cells and independently discovered their ability to trigger immune responses.

56. Inaba, K., Metlay, J. P., Crowley, M. T., & Steinman, R. M., 'Dendritic cells pulsed with protein antigens in vitro can prime antigen-specific, MHC-restricted T cells in situ', *The Journal of Experimental Medicine* 172, 631–40 (1990).

57. In 2013, one in seven researchers in Japan was female, compared to more than one in three in the UK, for example, as reported in 'Strengthening Japan's Research Capacity', from Kyoto University Gender Equality Promotion Center, available online here: http://www.cwr.kyoto-u.ac.jp/english/introduction.php. The original source of these data is quoted in this report (but available only in Japanese): http://www.stat.go.jp/data/kagaku/kekka/topics/topics80.htm. The topic is also discussed in the *Japan Times* here: http://www.japantimes.co.jp/news/2014/04/15/national/japans-scientists-just-14-female/#.VZ5fmcvbJaQ.

58. Palucka, K., & Banchereau, J., 'Cancer immunotherapy via dendritic cells', *Nature Reviews Cancer* 12, 265–77 (2012).

59. Engber, D., 'Is the cure for cancer inside you?', *New York Times Magazine*, 21 December 2012.

60. The problem is that if you accidently get exposed to your own blood cells that have been treated in some way, they could cause problems, whereas blood cells from somebody else would normally be destroyed on account of the genetic differences, like in a transplant situation.

61. Steenhuysen, J., & Nichols, M., 'Insight: Nobel winner's last big experiment: Himself', *Reuters*, 6 October 2011.

62. Engber.

63. Gravitz, L., 'A fight for life that united a field', *Nature* **478**, 163–4 (2011).

64. Steenhuysen & Nichols.

65. Gravitz.

66. Steenhuysen & Nichols.

67. Engber.

68. Steenhuysen & Nichols.

69. Engber.

70. Ibid.

71. Steinman (2011).

72. Interview with Andrew MacDonald, 24 August 2015.

73. Tirrell et al.

74. Palucka & Banchereau.

75. Interview with Christian Münz, 28 August 2015.

76. Ibid.

Chapter Three

1. Bresalier, M., '80 years ago today: MRC researchers discover viral cause of flu', *Guardian*, 8 July 2013.

2. I wrote about Macfarlane Burnet's life and work in detail in Chapter Two of *The Compatibility Gene* (Allen Lane, 2013).

3. Watts, G., 'Jean Lindenmann', *Lancet* **385**, 850 (2015).

4. Ibid.

5. There are lots of places where any scientific journey could be said to have begun: much started with Aristotle, or Darwin, for example. There were hints of cytokines in experiments performed before Lindenmann and Isaacs but because of the depth of their work, and the lucidity of their ideas, they are widely regarded as discoverers of the first cytokine.

6. Andrewes, C. H., 'Alick Isaacs. 1921–1967', *Biographical Memoirs of Fellows of the Royal Society* **13**, 205–21 (1967).

7. Edelhart, M., *Interferon: The New Hope for Cancer* (Orbis, 1982).

8. Findlay, G. M., & MacCallum, F. O., 'An interference phenomenon in relation to yellow fever and other viruses', *Journal of Pathology and Bacteriology* **44**, 405–24 (1937).

9. In detail, they used a virus that was inactivated by heat, so that it couldn't replicate on its own. And they used so-called red blood cell ghosts, which are red blood cells with the haemoglobin removed, so that the cells appear clearer in electron microscope pictures.

10. Pieters, T., *Interferon: The Science and Selling of a Miracle Drug* (Routledge, 2005).

11. Lindenmann, J., 'Preface' in Edelhart.

12. Isaacs, A., & Lindenmann, J., 'Virus interference. I. The interferon', *Proceedings of the Royal Society of London. Series B, Biological sciences* **147**, 258–67 (1957); Isaacs, A., Lindenmann, J., & Valentine, R. C., 'Virus interference. II. Some properties of interferon', *Proceedings of the Royal Society of London. Series B, Biological sciences* **147**, 268–73 (1957).

13. Pieters.

14. One eminent US scientist who doubted Lindenmann's and Isaacs' early evidence for interferon was Howard Temin. An expert in viruses, he went on to share the 1975 Nobel Prize in Physiology or Medicine – with Renato Dulbecco and David Baltimore – for the discovery of reverse transcriptase, a crucial enzyme for many viruses, e.g. HIV. This enzyme refuted a long-standing dogma that information encoded on DNA can be transfered to RNA, but not the other way around. Just like interferon, many scientists doubted the discovery of reverse transcriptase at first.

15. Pieters.

16. Edelhart.

17. Pieters.

18. Hall, S. S., *A Commotion in the Blood: Life, Death, and the Immune System* (Henry Holt and Company, 1997).

19. Ibid.

20. Interview with Leslie Brent, 23 October 2015.

21. Brent, L., 'Susanna Isaacs Elmhirst obituary', *Guardian*, 29 April 2010.

22. Hall.

23. Pieters.

24. Hall.

25. Isaacs, A., & Burke, D. C., 'Interferon: A possible check to virus infections', *New Scientist* **4**, 109–11 (1958).

26. Derek Burke recounts the story of interferon in an online article, 'The Discovery of Interferon, the First Cytokine, by Alick Isaacs and Jean Lindenmann in 1957', posted on 14 February 2009, available here: http://brainimmune.com/the-discovery-of-interferon-the-first-cytokine-by-alick-isaacs-and-jean-lindenmann-in-1957/.

27. Pieters.

28. Andrewes.

29. Cantell, K., *The Story of Interferon: The Ups and Downs in the Life of a Scientist* (World Scientific Publishing Co., 1998).

30. Hall.

31. Only a few viruses can sometimes cause cancer in humans, and these include human papilloma virus (HPV), Epstein–Barr virus (EBV) and human T-lymphotropic virus 1 (HTLV-1). Most people infected with these viruses do not develop cancer.

32. Gresser, I., & Bourali, C., 'Exogenous interferon and inducers of interferon in the treatment Balb-c mice inoculated with RC19 tumour cells', *Nature* **223**, 844–5 (1969).

33. Hall.

34. Gresser, I., 'Production of interferon by suspensions of human leucocytes', *Proceedings of the Society for Experimental Biology and Medicine* **108**, 799–803 (1961).

35. Cantell.

36. Ibid.

37. Ibid.

38. Pieters.

39. Cantell.

40. At one time, Cantell was flown to Cuba and got to meet Fidel Castro, who set up a Cuban research institute focused on interferon.

41. 'The Big IF in Cancer', *Time*, 31 March 1980.

42. Interview with Jordan Gutterman, 18 January 2016.

43. Gutterman and Lasker would chat on the phone three or four times a week. Once, Lasker phoned Gutterman at around midnight and she wanted to know about prostate cancer. Gutterman was asleep but stirred to answer the phone. 'Tell me what you know about prostate cancer,' Lasker said. Half asleep, Gutterman simply replied: 'It's more common in men.'

44. Cantell.
45. The chief reason that interferon was hard to purify was that it was only secreted from cells in tiny amounts. This is a characteristic of all cytokines, and that very small quantities still have strong effects in the body.
46. Cantell.
47. Ibid.
48. Herbert Boyer and Stanley Cohen first reported bacteria with genetic information from a different species in 1973. In their case, frog DNA was inserted into bacteria. Before this, in 1972, Paul Berg had combined DNA from different species into what is termed recombinant DNA molecules. Boyer was one of the co-founders of the biotech company Genentech.
49. Human insulin, initially sold as Humulin, was manufactored by Eli Lilly under licence from Genentech. FDA approval was given five months after the application was made, instead of the normal twenty to thirty months; see Lawrence Altman, 'A new insulin given approval for use in US', *New York Times*, 30 October 1982.
50. This uses a type of enzyme called reverse transcriptase, which can convert RNA into DNA, discovered by Howard Temin at the University of Wisconsin–Madison and, independently, by David Baltimore at MIT; the same Howard Temin who had doubted the existence of interferon at first.
51. Nagata, S., et al., 'Synthesis in E. coli of a polypeptide with human leukocyte interferon activity', *Nature* **284**, 316–20 (1980).
52. Cantell.
53. Other co-founders of Biogen included Phillip Sharp from MIT, and Nobel laureate Walter Gilbert from Harvard University.
54. Cantell.
55. Ibid.
56. 'The Big IF in Cancer', *Time*, 31 March 1980.
57. Panem, S., *The Interferon Crusade* (Brookings Institution, 1984).
58. Dickson, D., 'Deaths halt interferon trials in France', *Science* **218**, 772 (1982).
59. Panem.
60. Ahmed, S., & Rai, K. R., 'Interferon in the treatment of hairy-cell leukemia', *Best Practice and Research Clinical Haematology* **16**, 69–81 (2003).

61. Taniguchi, T., Fujii-Kuriyama, Y., & Muramatsu, M., 'Molecular cloning of human interferon cDNA', *Proceedings of the National Academy of Sciences USA* **77**, 4003–6 (1980).

62. Sorg, C., 'Lymphokines, monokines, cytokines', *Chemical Immunology* **49**, 82–9 (1990).

63. Interview with Werner Müller, 11 January 2016.

64. Atwood, M., *Moral Disorder* (Bloomsbury, 2006).

65. Cytokines are hormones in the sense that they are soluble factors produced by one cell which then affect the behaviour of other cells. Characteristics of some cytokines, however, are not usually attributed to hormones: some cytokines act relatively locally in the body; for example, a few are bound to the surface of cells rather than released into the bulk liquid environment, and some cytokines can be produced by many different types of cells.

66. McNab, F., Mayer-Barber, K., Sher, A., Wack, A., & O'Garra, A., 'Type I interferons in infectious disease', *Nature Reviews Immunology* **15**, 87–103 (2015).

67. Yan, N., & Chen, Z. J., 'Intrinsic antiviral immunity', *Nature Immunology* **13**, 214–22 (2012).

68. Everitt, A. R., et al., 'IFITM3 restricts the morbidity and mortality associated with influenza', *Nature* **484**, 519–23 (2012).

69. Ibid.

70. Zhang, Y. H., et al., 'Interferon-induced transmembrane protein-3 genetic variant rs12252–C is associated with severe influenza in Chinese individuals', *Nature Communications* **4**, 1418 (2013).

71. Ibid.

72. Interview with Peter Openshaw, 5 January 2016.

73. Chesarino, N. M., McMichael, T. M., & Yount, J. S., 'E3 Ubiquitin Ligase NEDD4 Promotes Influenza Virus Infection by Decreasing Levels of the Antiviral Protein IFITM3', *PLoS Pathogens* **11**, e1005095 (2015).

74. If the same enzyme was targeted in humans, there could be side effects because this type of enzyme is involved in the degradation of many other protein molecules, not just the one encoded for in the IFITM3 gene.

75. Obviously, any statements I make regarding medical treatments should not be taken as explicit advice for one's own medical care. This book describes general principles and ideas which hopefully help one's understanding of the science behind health issues, but

this is not a substitute for specific advice from a physician or general practitioner. More information about the use of interferon in cancer therapies is available here, from Cancer Research UK: http://www.cancerresearchuk.org/about-cancer/cancers-in-general/treatment/cancer-drugs/interferon.

76. Zitvogel, L., Galluzzi, L., Kepp, O., Smyth, M. J., & Kroemer, G., 'Type I interferons in anticancer immunity', *Nature Reviews Immunology* **15**, 405–14 (2015).

77. Ibid.

78. Red blood cells are an exception; they are the only cells in our body that can't produce or respond to a cytokine.

79. Email correspondence with Salim Khakoo, 2 February 2017.

80. Rusinova, I., et al., 'Interferome v2.0: an updated database of annotated interferon-regulated genes', *Nucleic Acids Research* **41**, D1040–6 (2013).

81. The stories of how many cytokines were discovered are discussed in detail across several chapters in Smith, K. A. (ed.), *A Living History of Immunology. Frontiers in Immunology* **6**, 502 (2015).

82. The interleukins have been discovered at a rate of about one per year over the last decade, and we were at IL-23 in 2001, for example.

83. Dinarello, C. A., 'Immunological and inflammatory functions of the interleukin-1 family', *Annual Review of Immunology* **27**, 519–50 (2009).

84. You may have come across the importance of neutrophils if you or someone you know has had neutropenia. Caused by cancer and some cancer therapies, neutropenia is essentially having low numbers of neutrophils, making a person more susceptible to infections.

85. Brinkmann, V., et al., 'Neutrophil extracellular traps kill bacteria', *Science* **303**, 1532–5 (2004).

86. Kolaczkowska, E., & Kubes, P., 'Neutrophil recruitment and function in health and inflammation', *Nature Reviews Immunology* **13**, 159–75 (2013).

87. IL-2 has been important in thousands of research studies; in Robert Gallo's lab at the National Institutes of Health, for example, it allowed the isolation of HIV from T cells.

88. Howard, M., & O'Garra, A., 'Biological properties of interleukin 10', *Immunology Today* **13**, 198–200 (1992).

89. Kuhn, R., Lohler, J., Rennick, D., Rajewsky, K., & Muller, W., 'Interleukin-10–deficient mice develop chronic enterocolitis', *Cell* **75**, 263–74 (1993).

90. The charity Crohn's and Colitis UK provides details on these diseases: http://www.crohnsandcolitis.org.uk/

91. 'Partnership for Public Service, Dr Steven Rosenberg: Saving lives through important breakthroughs in cancer treatment', *Washington Post*, 6 May 2015.

92. Fox, T., 'The federal employee of the year', *Washington Post*, 7 October 2015.

93. Rosenberg's high status is emphasised by the fact that in 1985, he was the cancer expert on the surgical team who operated on President Reagan. At a press conference, it was Rosenberg who said the words which became the headline for newspapers around the world: 'The president has cancer'. Others on the team presented the details of the president's condition, but Rosenberg thought it important to use the word cancer, to demystify the word itself and to make the disease less isolating.

94. Hall.

95. Rosenberg, S. A., & Barry, J. M., *The Transformed Cell: Unlocking the Mysteries of Cancer* (Orion, 1992).

96. Ibid.

97. Rosenberg goes on to say: 'Now am I going to achieve that goal, I think it's very unlikely. But can I put a dent into that goal, I think so.' He says this in an interview in 2007, for Roadtrip Nation, a series of videos in which young people interview people who made careers from things they love. A clip from Rosenberg's interview is available online here: https://www.youtube.com/watch?v=iNc_nY6nUoI.

98. Rosenberg & Barry.

99. It is possible that small numbers of cancer cells are dealt with more commonly by our immune system without us being aware of it – and we're not aware of it.

100. Rosenberg & Barry.

101. Ibid.

102. Ibid.

103. Hall.

104. In all the scientific reports discussing this case, Linda Taylor is given the pseudonym Linda Granger, to protect her identity,

but she is named as Linda Taylor in some more recent presentations of her case, including the 2015 PBS TV documentary *Cancer: The Emperor of all Maladies*.

105. Rosenberg & Barry.
106. Rosenberg, S. A., et al., 'Observations on the systemic administration of autologous lymphokine-activated killer cells and recombinant interleukin-2 to patients with metastatic cancer', *New England Journal of Medicine* **313**, 1485–92 (1985).
107. Rosenberg & Barry.
108. Before Rosenberg knew that Taylor's treatment had worked, he also treated another patient, given the pseudonym James Jensen. Rosenberg says in his autobiography that he 'felt desperate, as desperate as I have ever felt in my life', and he pushed Jensen even harder than he did Taylor, giving him even more infusions of immune cells as well as high doses of IL-2. Jensen's tumours never disappeared, and he eventually died from his cancer, but his tumours did shrink in response to Rosenberg's effort.
109. Rosenberg was on the front cover of *Newsweek* on 19 December 1985. His success was also the lead news item on NBC and ABC news on TV, and front-page news in newspapers across the USA, Europe, China and Japan.
110. Schmeck, H. M., Jr, 'Cautious optimism is voiced about test cancer therapy', *New York Times*, 6 December 1985.
111. Rosenberg & Barry.
112. Burns, K., *Cancer: The Emperor of All Maladies* (PBS TV, 2015).
113. Rosenberg, S. A., 'IL-2: the first effective immunotherapy for human cancer', *The Journal of Immunology* **192**, 5451–8 (2014).
114. Rosenberg & Barry.
115. Ibid.
116. Coventry, B. J., & Ashdown, M. L., 'The 20th anniversary of interleukin-2 therapy: bimodal role explaining longstanding random induction of complete clinical responses', *Cancer Management and Research* **4**, 215–21 (2012).

Chapter Four

1. Feldmann, M., 'Translating molecular insights in autoimmunity into effective therapy', *Annual Review of Immunology* **27**, 1–27 (2009).
2. Ibid.

3. Interview with Marc Feldmann, 22 February 2016.

4. Feldmann (2009).

5. Interview with Marc Feldmann, 22 February 2016.

6. Interview with Werner Müller, 11 January 2016.

7. Dinarello, C. A., 'Historical insights into cytokines', *European Journal of Immunology* **37 Suppl** 1, S34–45 (2007).

8. Auron, P. E., et al., 'Nucleotide sequence of human monocyte interleukin 1 precursor cDNA', *Proceedings of the National Academy of Sciences USA* **81**, 7907–11 (1984).

9. Email correspondence with Werner Müller, 13 January 2016.

10. Lachman, L. B., 'Summary of the Fourth International Lymphokine Workshop', *Lymphokine Research* **4**, 51–7 (1985).

11. Interview with Werner Müller, 11 January 2016.

12. Ibid.

13. Lachman.

14. Gannes, S., 'Striking it rich in biotech', *Fortune Magazine*, 9 November 1987. Available online here: http://archive.fortune.com/magazines/fortune/fortune_archive/1987/11/09/69810/index.htm.

15. Interview with Werner Müller, 11 January 2016.

16. Lachman.

17. March, C. J., et al., 'Cloning, sequence and expression of two distinct human interleukin-1 complementary DNAs', *Nature* **315**, 641–7 (1985).

18. Wolff, S. M., et al., 'Clone controversy at Immunex', *Nature* **319**, 270 (1986).

19. Marshall, E., 'Battle ends in $21 million settlement', *Science* **274**, 911 (1996).

20. Ibid.

21. Ibid.

22. 'Immunex to Pay $21 Million To Cistron to Settle Lawsuit', *Wall Street Journal*, 4 November 1996. Available online here: http://www.wsj.com/articles/SB847060346541962500.

23. Marshall.

24. Hamilton, D. P., 'Amgen Confirms Cash, Stock Deal to Acquire Smaller Rival Immunex', *Wall Street Journal*, 18 December 2001. Available online here: http://www.wsj.com/articles/SB1008606575817774000.

25. Feldmann (2009).

26. Bottazzo, G. F., Pujol-Borrell, R., Hanafusa, T., & Feldmann, M., 'Role of aberrant HLA-DR expression and antigen presentation in induction of endocrine autoimmunity', *Lancet* **2**, 1115–19 (1983).

27. Interview with Marc Feldmann, 22 February 2016.

28. Data on the prevalence of rheumatoid arthritis is available from the Centers for Disease Control and Prevention, a part of the US Department of Health and Human Services, online here http://www.cdc.gov/arthritis/basics/rheumatoid.htm.

29. Eyre, S., et al., 'High-density genetic mapping identifies new susceptibility loci for rheumatoid arthritis', *Nature Genetics* **44**, 1336–40 (2012).

30. Heliovaara, M., et al., 'Coffee consumption, rheumatoid factor, and the risk of rheumatoid arthritis', *Annals of the Rheumatic Diseases* **59**, 631–5 (2000).

31. Lee, Y. H., Bae, S. C., & Song, G. G., 'Coffee or tea consumption and the risk of rheumatoid arthritis: a meta-analysis', *Clinical Rheumatology* **33**, 1575–83 (2014).

32. Feldmann (2009).

33. Uganda became independent in 1962.

34. 'Spotlight: Ravinder Maini – A Career in Research. An interview with Ravinder Maini' published on 28 March 2014, in three parts: the first part is available online here: https://www.youtube.com/watch?v=ZJ53ApfoiD8.

35. Interview with Marc Feldmann, 22 February 2016.

36. Interview with Ravinder Maini, 15 February 2016.

37. Feldmann and Maini weren't the only ones doing this kind of research at the time. Several different research groups across the globe had set out to determine the cytokines present in a diseased joint, often using different methods.

38. TNF-α should have a simpler name, just another numbered inter-leukin, IL-something, but just like interferon, it was discovered and named before the interleukin classification was in place and its cumbersome name has stuck.

39. Buchan, G., et al., 'Interleukin-1 and tumour necrosis factor mRNA expression in rheumatoid arthritis: prolonged production of IL-1 alpha', *Clinical and Experimental Immunology* **73**, 449–55 (1988).

40. Carswell, E. A., et al., 'An endotoxin-induced serum factor that causes necrosis of tumors', *Proceedings of the National Academy of Sciences USA* **72**, 3666–70 (1975).

41. Vilček, J., *Love and Science: A Memoir* (Seven Stories Press, 2016).

42. In 1938, with the security of Jews in Czechoslovakia already in doubt, his family hatched a plan to leave young Vilček with a family in Holland. The plan never materialised. If it had done, Vilček would have suffered the same fate as the Jewish Dutch family he would have been staying with, and they did not survive the Holocaust.

43. While the nuns could not have been Nazi sympathisers – they helped Jewish children – government regulations for the orphanage required Vilček to be taught loyalty to the Nazis.

44. Vilček (2016).

45. Ibid.

46. This was a view advocated by Stalin's confidant Trofim Lysenko, who claimed that seeds treated with moisture and cold grew better in the Russian frost, and that this benefit passed on to future generations of seeds. Lysenko's theories are nowadays widely regarded as wrong. Scientists who produced, or faked, evidence to support his ideas were favoured with rewards, recognition and research funding. There is, however, relatively recent evidence that some genetic changes caused by the environment can be passed on from one generation to the next; the emerging field of epigenetics. This does not mean that Lysenko's ideas were right, because the effects are small and limited. Scientist and broadcaster Adam Rutherford warns against over-hyping epigenetics to a mystical level, in 'Beware the pseudo gene genies', *Guardian*, 19 July 2015. Available online here: https://www.theguardian.com/science/2015/jul/19/epigenetics-dna–darwin-adam-rutherford.

47. Vilček, J., 'From IFN to TNF: a journey into realms of lore', *Nature Immunology* 10, 555–7 (2009). In this article, Vilček recalls meeting Isaacs in 1958. But in discussion with him on 4 February 2016, he now thinks it must have been in 1957, as that was the year he graduated in Bratislava.

48. Interview with Jan Vilček, 4 February 2016.

49. Vilček, J., 'An interferon-like substance released from tickborne encephalitis virus-infected chick embryo fibroblast cells', *Nature* 187, 73–4 (1960).

50. He didn't like the restrictions on travel that the communist country imposed on him, and had had been called in for a few

interviews with the secret police, but says it was his wife who really pushed him to defect.

51. Interview with Jan Vilček, 4 February 2016.

52. Perez-Penauug, R., 'Research Scientist Gives $105 Million to NYU', *New York Times*, 12 August 2005.

53. Interview with Jan Vilček, 4 February 2016.

54. When winning the prestigious Albert Lasker Basic Medical Research Award for this work, César Milstein and Georges Köhler publicly stated that 'both the conception and execution of the work was the result of close collaboration between us with the skilled technical assistance of Shirley Howe'.

55. Köhler, G., & Milstein, C., 'Continuous cultures of fused cells secreting antibody of predefined specificity', *Nature* **256**, 495–7 (1975).

56. Margulies, D. H., 'Monoclonal antibodies: producing magic bullets by somatic cell hybridization', *The Journal of Immunology* **174**, 2451–2 (2005).

57. The original agreement with Centocor was based on Vilček producing an antibody against interferon which the company planned to develop into a diagnostic test for levels of interferon in, for example, a blood sample.

58. Vilček was very surprised when he first arrived in New York that he had to obtain all of his own funding to do research. In communist Czechoslovakia, there were many shortages of equipment and suchlike, but each researcher was at least given money for their research, without having to write specific grant applications.

59. With Vilček, Junming Le also shared in the royalties that arose from the anti-TNFα antibody. Le set up the Iris and Junming Le Foundation in 2006, to support various medical and health activities.

60. Marks, L. V., *The Lock and Key of Medicine: Monoclonal Antibodies and the Transformation of Healthcare* (Yale University Press, 2015).

61. Beutler, B., Milsark, I. W., & Cerami, A. C., 'Passive immunization against cachectin/tumor necrosis factor protects mice from lethal effect of endotoxin', *Science* **229**, 869–71 (1985).

62. Lagu, T., et al., 'Hospitalizations, costs, and outcomes of severe sepsis in the United States 2003 to 2007', *Critical Care Medicine* **40**, 754–61 (2012).

63. Marks, L., 'The birth pangs of monoclonal antibody therapeutics: the failure and legacy of Centoxin', *mAbs* **4**, 403–12 (2012).

64. John Ghrayeb and his team at Centocor did this work, using methods established in 1983–5 by several research teams working independently, involving academic labs in Cambridge (UK), Stanford and Toronto, as well as the US company Becton-Dickinson in California.

65. Vilček (2009).

66. Ibid.

67. Brennan, F. M., Chantry, D., Jackson, A., Maini, R., & Feldmann, M., 'Inhibitory effect of TNF alpha antibodies on synovial cell interleukin-1 production in rheumatoid arthritis', *Lancet* **2**, 244–7 (1989).

68. Interview with Marc Feldmann, 22 February 2016.

69. Williams, R. O., Feldmann, M., & Maini, R. N., 'Anti-tumor necrosis factor ameliorates joint disease in murine collagen-induced arthritis', *Proceedings of the National Academy of Sciences USA* **89**, 9784–8 (1992).

70. As well as work that Feldmann was directly involved with, another line of evidence in support of his ideas came from George Kollias and his colleagues in Greece. They showed that mice genetically engineered to produce human TNFα would develop an inflammation in their joints, consistent with the idea that this cytokine was very important in arthritis.

71. Feldmann (2009).

72. Interview with Jan Vilček, 4 February 2016.

73. Interview with Ravinder Maini, 15 February 2016.

74. Ibid.

75. Feldmann (2009).

76. Feldmann, M., 'Development of anti-TNF therapy for rheumatoid arthritis', *Nature Reviews Immunology* **2**, 364–71 (2002).

77. Interview with Jan Vilček, 4 February 2016.

78. Interview with Ravinder Maini, 15 February 2016.

79. Feldmann (2009).

80. Ibid.

81. Interview with Marc Feldmann, 22 February 2016.

82. Vilček (2016).

83. The company obtained approval to use Remicade first for Crohn's disease, a chronic inflammation in the intestine, and then later for rheumatoid arthritis. No new treatments for Crohn's disease had been approved for decades. So Crohn's disease qualified as being

a relatively rare disease with few treatment options, which meant that fewer people were required to be tested in a clinical trial before the drug was approved. The paucity of treatments also meant that regulatory review of the drug was prioritised. Overall, this meant that the cost of getting the drug approved for Crohn's disease was far less than getting it approved for rheumatoid arthritis.

84. Morrow, D. J., 'Johnson & Johnson to Acquire Centocor', *New York Times*, 22 July 1999.

85. Vilček (2016).

86. Feldmann (2009).

87. Feldmann (2002).

88. There are several ways in which fully human antibodies can be made. One way is to use mice which have been modified so that their antibody-making genes have been replaced with human genes. However, Humira, a fully human anti-TNFα antibody approved in the USA in 2002, was made in a different way, using viruses that invade bacteria, by a technique called phage display.

89. Number of users taken from the official website for Remicade: http://www.remicade.com/.

90. White, E. B., *Here is New York* (Harper & Bros., 1949).

91. Vilček (2016).

92. Interview with Ravinder Maini, 15 February 2016.

93. Choy, E. H., Kavanaugh, A. F., & Jones, S. A., 'The problem of choice: current biologic agents and future prospects in RA', *Nature Reviews Rheumatology* **9**, 154–63 (2013).

94. Winthrop, K. L., & Chiller, T., 'Preventing and treating biologic-associated opportunistic infections', *Nature Reviews Rheumatology* **5**, 405–10 (2009).

95. Choy et al.

96. 'Spotlight: Ravinder Maini – A Career in Research. An interview with Ravinder Maini' published on 28 March 2014, in three parts: the first part is available online here: https://www.youtube.com/watch?v=ZJ53ApfoiD8.

97. Another reason that this may happen for some versions of anti-TNFα therapy is that the body begins to produce antibodies against the therapeutic antibody, which stop it from working.

98. Occasionally, patients with Crohn's disease have come off treatment with anti-TNFα therapy and not relapsed, but this is rare and the reasons are not understood.

99. Marks (2015).

100. The last line of Milstein's and Köhler's paper on the culture of hybridoma cells for the production of monoclonal antibodies, published in *Nature* in 1975, simply reads: 'Such cultures could be valuable for medical and industrial use.' In its simplicity, this is reminiscent of another classic ending to a scientific paper: Watson's and Crick's famous line in their description of the double helix structure of DNA, published in *Nature* in 1973, which reads: 'It has not escaped our notice that the specific pairing we have postulated immediately suggests a possible copying mechanism for the genetic material.'

101. Dorner, T., Radbruch, A., & Burmester, G. R., 'B-cell-directed therapies for autoimmune disease', *Nature Reviews Rheumatology* **5**, 433–41 (2009).

102. The World Health Organization's list of essential medicines can be downloaded here: http://www.who.int/medicines/services/essmedicines_def/en/.

103. Battella, S., Cox, M. C., Santoni, A., & Palmieri, G., 'Natural killer (NK) cells and anti-tumor therapeutic mAb: unexplored interactions', *Journal of Leukocyte Biology* **99**, 87–96 (2016).

104. Rudnicka, D., et al., 'Rituximab causes a polarization of B cells that augments its therapeutic function in NK-cell-mediated antibody-dependent cellular cytotoxicity', *Blood* **121**, 4694–702 (2013).

105. 'Drug trial victim's "hell" months', BBC News online, http://news.bbc.co.uk/1/hi/health/5121824.stm.

106. Vince, G., 'UK drug trial disaster – the official report', *New Scientist*, 25 May 2006.

107. Horvath, C. J., & Milton, M. N., 'The TeGenero incident and the Duff Report conclusions: a series of unfortunate events or an avoidable event?', *Toxicologic Pathology* **37**, 372–83 (2009).

Chapter Five

1. Strominger, J. L., 'The tortuous journey of a biochemist to immunoland and what he found there', *Annual Review of Immunology* **24**, 1–31 (2006).

2. Bauer, S., et al., 'Activation of NK cells and T cells by NKG2D, a receptor for stress-inducible MICA', *Science* **285**, 727–9 (1999).

3. Van der Zee, J., 'Heating the patient: a promising approach?', *Annals of Oncology* **13**, 1173–84 (2002).

4. Shen, R. N., Hornback, N. B., Shidnia, H., Shupe, R. E., & Brahmi, Z., 'Whole-body hyperthermia decreases lung metastases in lung tumor-bearing mice, possibly via a mechanism involving natural killer cells', *Journal of Clinical Immunology* **7**, 246–53 (1987).

5. Kokolus, K. M., et al., 'Baseline tumor growth and immune control in laboratory mice are significantly influenced by subthermo-neutral housing temperature', *Proceedings of the National Academy of Sciences USA* **110**, 20176–81 (2013).

6. Evans, S. S., Repasky, E. A., & Fisher, D. T., 'Fever and the thermal regulation of immunity: the immune system feels the heat', *Nature Reviews Immunology* **15**, 335–49 (2015).

7. Elinav, E., et al., 'Inflammation-induced cancer: crosstalk between tumours, immune cells and microorganisms', *Nature Reviews Cancer* **13**, 759–71 (2013).

8. Zelenay, S., et al., 'Cyclooxygenase-Dependent Tumor Growth through Evasion of Immunity', *Cell* **162**, 1257–70 (2015).

9. Groh, V., Wu, J., Yee, C., & Spies, T., 'Tumour-derived soluble MIC ligands impair expression of NKG2D and T-cell activation', *Nature* **419**, 734–8 (2002).

10. Deng, W., et al., 'Antitumor immunity. A shed NKG2D ligand that promotes natural killer cell activation and tumor rejection', *Science* **348**, 136–9 (2015).

11. Evans et al.

12. Ibid.

13. Ibid.

14. Lafrance, A., 'A cultural history of the fever', *Atlantic*, 16 September 2015.

15. Evans et al.

16. Rice, P., et al., 'Febrile-range hyperthermia augments neutrophil accumulation and enhances lung injury in experimental gram-negative bacterial pneumonia', *The Journal of Immunology* **174**, 3676–85 (2005).

17. In some situations a fever warrants medical attention, such as a fever in a newborn baby. There is plenty of advice on the Internet about fevers but stick to web pages, such as those from the National Health Service in the UK, which present the consensus scientific view.

18. Woolf, V., *On Being Ill* (Hogarth Press, 1930).

19. Interview with Luke O'Neill, 16 March 2016.

20. Kalinski, P., 'Regulation of immune responses by prostaglandin E2', *The Journal of Immunology* **188**, 21–8 (2012).

21. Furuyashiki, T., & Narumiya, S., 'Stress responses: the contribution of prostaglandin E(2) and its receptors', *Nature Reviews Endocrinology* **7**, 163–75 (2011).

22. According to the International Aspirin Foundation, aspirin is one of the mostly widely used pharmaceutical drugs in the world, with 100 billion tablets'-worth being produced every year. The story of aspirin is, of course, fascinating in its own right, worthy of a whole book. Briefly, the active ingredient of aspirin is a synthetic version of a chemical found in the willow tree (and other plants). In 1823, a form of aspirin was isolated from willow bark. The pharmaceutical company Bayer made a synthetic version in 1897 and clinical trials began. Aspirin was launched in 1899 and was arguably the first ever drug produced from the pharmaceutical industry. In the 1930s, Bayer's patents ran out and aspirin became a generic drug. Beginning in 1969, a series of experiments by John Vane working in London, UK, established that aspirin inhibited the production of prostaglandins. Vane tells the story in his Nobel Lecture from 1982, online here: http://www.nobelprize.org/mediaplayer/index.php?id=1615

23. Slocumb, C. H., 'Philip Showalter Hench, 1896–1965. In Memoriam', *Arthritis and Rheumatism* **8**, 573–6 (1965).

24. Hench, P. S., 'The reversibility of certain rheumatic and non-rheumatic conditions by the use of cortisone or of the pituitary adrenocorticotropic hormone' (Nobel Lecture, 1950) in *Nobel Lectures, Physiology or Medicine 1942–1962* (Elsevier, 1964).

25. Tata, J. R., 'One hundred years of hormones', *EMBO Reports* **6**, 490–6 (2005).

26. Reichstein also worked out a way to synthesise vitamin C, which helped lead to its mass production. Kendall also isolated hormones from the thyroid, including thyroxine, sometimes known as T4.

27. Reichstein, T., 'Chemistry of the Adrenal Cortex Hormones' (Nobel Lecture, 1950) in *Nobel Lectures, Physiology or Medicine 1942–1962* (Elsevier, 1964).

28. Kendall, E. C., *Cortisone: Memoirs of a Hormone Hunter* (Charles Scribner's Sons, 1971).

29. Rooke, T., *The Quest for Cortisone* (Michigan State University Press, 2012).

30. Hench.

31. Saenger, A. K., 'Discovery of the wonder drug: from cows to cortisone. The effects of the adrenal cortical hormone 17–hydroxy-11–dehydrocorticosterone (Compound E) on the acute phase of rheumatic fever; preliminary report. *Mayo Clinic Proceedings* 1949;24:277–97', *Clinical Chemistry* **56**, 1349–50 (2010).

32. Le Fanu, J., *The Rise and Fall of Modern Medicine* (revised edition, Abacus, 2011).

33. Rooke.

34. Ibid.

35. Ibid.

36. Hench.

37. Rooke.

38. Le Fanu.

39. Rooke.

40. Le Fanu.

41. Cortisol is a steroid hormone in the glucocorticoid family, not to be confused with anabolic steroids which are different and are related to the male sex hormone testosterone. Anabolic steroids increase muscle growth and are sometimes illegally used in body-building and sports. Anabolic steroids can also be used therapeutically, to stimulate muscle growth in chronic wasting diseases and in some cases, to help treat breast cancer.

42. Chrousos, G. P., 'Stress and disorders of the stress system', *Nature Reviews Endocrinology* **5**, 374–81 (2009).

43. Both cortisone and cortisol are produced as hormones by the adrenal gland in the body: the addition of a hydrogen atom converts cortisone to cortisol, and cortisol is the more potent, or active, version of the hormone.

44. Maisel, A. Q., *The Hormone Quest* (Random House, 1965).

45. 'MRC Streptomycin in Tuberculosis Trials Committee. Streptomycin treatment of pulmonary tuberculosis', *British Medical Journal* **2**, 769–82 (1948).

46. Interview with David Ray, 15 April 2016.

47. Paget, S. A., Lockshin, M. D., & Loebl, S., *The Hospital for Special Surgery Rheumatoid Arthritis Handbook* (John Wiley & Sons, 2002).

48. High levels of cortisol can induce symptoms known as Cushing's syndrome. As well as muscle weakness, fatigue and weight gain, these symptoms also can include thinning skin, reddish-purple marks on arms and legs, low libido and fat deposits in the face. As well as these symptoms developing in people taking cortisol or one of its derivatives as a medicine for a long period of time, they can also arise where a tumour develops in one of the body's glands that produce stress hormones.

49. Rooke.

50. The 153 boxes of items collected by Philip Hench about the history of yellow fever are stored at the University of Virginia. The archive also includes many personal items, such as photographs of Philip Hench with his wife Mary in Havana, and many personal letters and telegrams. Some items have been digitised and are available online here: https://search.lib.virginia.edu/catalog/uva-lib:2513789.

51. Kendall.

52. Selye, H., 'A syndrome produced by diverse nocuous agents', *Nature* **138** (1936).

53. Hans Selye was nominated for a Nobel Prize for many years, by at least seventeen different individuals, but he never won. Details archived here in the 'Nomination Database' at nobelprize.org. Nobel Media AB 2014. <http://www.nobelprize.org/nomination/archive/show_people.php?id=8395>

54. 'Obituary. Dr Hans Selye dies in Montreal; studied effects of stress on body', *New York Times*, 22 October 1982.

55. Fink, G., 'In retrospect: Eighty years of stress', *Nature* **539**, 175–6 (2016).

56. Selye, H., *The Stress of Life* (McGraw-Hill, 1956).

57. 'Obituary', *New York Times*, 22 October 1982.

58. Ibid.

59. There is great diversity in what the word 'stress' applies to, and although some aspects of these different stressors are the same, there are also differences. Stress due to divorce doesn't have the same impact on the body as, for example, stress due to being stuck in a traffic jam, and the situation is even more complicated because everyone responds differently to any given stressor. Nevertheless, there is common ground in the particular hormones involved in stress and because of this, a modern definition of stress includes

it being something that stimulates the so-called hypothalamic-pituitary-adrenal axis, which leads to the production of cortisol.

60. Gamble, K. L., Berry, R., Frank, S. J., & Young, M. E., 'Circadian clock control of endocrine factors', *Nature Reviews Endocrinology* **10**, 466–75 (2014).

61. Webster, J. I., Tonelli, L., & Sternberg, E. M., 'Neuroendocrine regulation of immunity', *Annual Review of Immunology* **20**, 125–63 (2002).

62. Ironson, G., et al., 'Posttraumatic stress symptoms, intrusive thoughts, loss, and immune function after Hurricane Andrew', *Psychosomatic Medicine* **59**, 128–41 (1997).

63. Padgett, D. A., & Glaser, R., 'How stress influences the immune response', *Trends in Immunology* **24**, 444–8 (2003).

64. I write this as fact, not in advocacy for animal experiments, which is a complex issue widely debated elsewhere.

65. Glaser, R., & Kiecolt-Glaser, J. K., 'Stress-induced immune dysfunction: implications for health', *Nature Reviews Immunology* **5**, 243–51 (2005).

66. Rodriguez-Galan, M. C., et al., 'Immunocompetence of macrophages in rats exposed to Candida albicans infection and stress', *American Journal of Physiology. Cell Physiology* **284**, C111–18 (2003).

67. Vedhara, K., et al., 'Chronic stress in elderly carers of dementia patients and antibody response to influenza vaccination', *Lancet* **353**, 627–31 (1999).

68. Leserman, J., et al., 'Progression to AIDS: the effects of stress, depressive symptoms, and social support', *Psychosomatic Medicine* **61**, 397–406 (1999).

69. Cole, S. W., Kemeny, M. E., Taylor, S. E., Visscher, B. R., & Fahey, J. L., 'Accelerated course of human immunodeficiency virus infection in gay men who conceal their homosexual identity', *Psychosomatic Medicine* **58**, 219–31 (1996).

70. Glaser & Kiecolt-Glaser.

71. Brod, S., Rattazzi, L., Piras, G., & D'Acquisto, F., '"As above, so below" examining the interplay between emotion and the immune system', *Immunology* **143**, 311–18 (2014).

72. Pesce, M., et al., 'Positive correlation between serum interleukin-1beta and state anger in rugby athletes', *Aggressive Behaviour* **39**, 141–8 (2013).

73. Hayashi, T., et al., 'Laughter up-regulates the genes related to NK cell activity in diabetes', *Biomedical Research* **28**, 281–5 (2007).

74. Not much is understood about laughter in general, a complex social interaction that we share with all mammals. See, for example, a TED talk by Sophie Scott on 'Why we laugh': https://www.ted.com/talks/sophie_scott_why_we_laugh?language=en.

75. Fransen, M., Nairn, L., Winstanley, J., Lam, P., & Edmonds, J., 'Physical activity for osteoarthritis management: a randomized controlled clinical trial evaluating hydrotherapy or Tai Chi classes', *Arthritis and Rheumatism* **57**, 407–14 (2007).

76. Yang, Y., et al., 'Effects of a traditional Taiji/Qigong curriculum on older adults' immune response to influenza vaccine', *Medicine and Sport Science* **52**, 64–76 (2008).

77. Ho, R. T., et al., 'The effect of t'ai chi exercise on immunity and infections: a systematic review of controlled trials', *Journal of Alternative and Complementary Medicine* **19**, 389–96 (2013).

78. Ibid.

79. Ibid.

80. Ibid.

81. Morgan, N., Irwin, M. R., Chung, M., & Wang, C., 'The effects of mind–body therapies on the immune system: meta analysis', *PLoS One* **9**, e100903 (2014).

82. The NIH and NHS discuss the effects of t'ai chi here: https://nccih.nih.gov/health/taichi/introduction.htm and http://www.nhs.uk/Livewell/fitness/Pages/taichi.aspx.

83. Bhattacharya, A., McCutcheon, E. P., Shvartz, E., & Greenleaf, J. E., 'Body acceleration distribution and O_2 uptake in humans during running and jumping', *Journal of Applied Physiology: Respiratory, Environmental and Exercise Physiology* **49**, 881–7 (1980).

84. Briskin, S., & LaBotz, M., 'Trampoline safety in childhood and adolescence', *Pediatrics* **130**, 774–9 (2012).

85. Saxon, W., 'Elvin Kabat, 85, Microbiologist Known for Work in Immunology', *New York Times*, 22 June 2000.

86. Wax, R., *A Mindfulness Guide for the Frazzled* (Penguin, 2016).

87. Goyal, M., et al., 'Meditation programs for psychological stress and well-being: a systematic review and meta-analysis', *JAMA Internal Medicine* **174**, 357–68 (2014).

88. Kuyken, W., et al., 'Effectiveness and cost-effectiveness of mindfulness-based cognitive therapy compared with maintenance

antidepressant treatment in the prevention of depressive relapse or recurrence (PREVENT): a randomised controlled trial', *Lancet* **386**, 63–73 (2015).

89. Pickert, K., 'The art of being mindful', *Time*, 3 February 2014.
90. Black, D. S., & Slavich, G. M., 'Mindfulness meditation and the immune system: a systematic review of randomized controlled trials', *Annals of the New York Academy of Sciences* (2016).
91. Ibid.
92. O'Leary, K., O'Neill, S., & Dockray, S., 'A systematic review of the effects of mindfulness interventions on cortisol', *Journal of Health Psychology* (2015).

Chapter Six

1. Loudon, A. S., 'Circadian biology: a 2.5 billion-year-old clock', *Current Biology* **22**, R570–1 (2012).
2. Cutolo, M., 'Chronobiology and the treatment of rheumatoid arthritis', *Current Opinion in Rheumatology* **24**, 312–18 (2012).
3. Foster, R. G., & Kreitzman, L., *The Rhythms of Life: The Biological Clocks That Control the Daily Lives of Every Living Thing* (Profile Books, 2004).
4. Folkard, S., Lombardi, D. A., & Spencer, M. B., 'Estimating the circadian rhythm in the risk of occupational injuries and accidents', *Chronobiology International* **23**, 1181–92 (2006).
5. Foster & Kreitzman (2004).
6. Foster, R. G., & Kreitzman, L., 'The rhythms of life: what your body clock means to you!', *Experimental Physiology* **99**, 599–606 (2014).
7. Wright, M. C., et al., 'Time of day effects on the incidence of anesthetic adverse events', *Quality & Safety in Health Care* **15**, 258–63 (2006).
8. Bellet, M. M., et al., 'Circadian clock regulates the host response to salmonella', *American Journal of Physiology. Cell Physiology* **110**, 9897–902 (2013).
9. Gibbs, J., et al., 'An epithelial circadian clock controls pulmonary inflammation and glucocorticoid action', *Nature Medicine* **20**, 919–26 (2014).
10. Scheiermann, C., Kunisaki, Y., & Frenette, P. S., 'Circadian control of the immune system', *Nature Reviews Immunology* **13**, 190–8 (2013).

11. Email correspondence with David Ray, 20 April 2016.

12. Email correspondence with Eleanor Riley, 27 May 2016, and Robert Sinden, 10–11 June 2016.

13. Sinden, R. E., Butcher, G. A., Billker, O., & Fleck, S. L., 'Regulation of infectivity of Plasmodium to the mosquito vector', *Advances in Parasitology* **38**, 53–117 (1996).

14. Interview with Andrew Loudon, 6 May 2016.

15. Till Roenneberg, speaking at the Euroscience Open Forum, 27 July 2016. Roenneberg is a leading sleep researcher at the Ludwig Maximilian University of Munich, and is also the author of *Internal time: Chronotypes, Social Jet Lag, and Why You're So Tired* (Harvard University Press, 2012).

16. Durrington, H. J., Farrow, S. N., Loudon, A. S., & Ray, D. W., 'The circadian clock and asthma', *Thorax* **69**, 90–2 (2014).

17. Foster & Kreitzman (2004).

18. Cutolo.

19. Foster & Kreitzman (2004).

20. Litinski, M., Scheer, F. A., & Shea, S. A., 'Influence of the Circadian System on Disease Severity', *Sleep Medicine Clinics* **4**, 143–63 (2009).

21. Filipski, E., et al., 'Effects of chronic jet lag on tumor progression in mice', *Cancer Research* **64**, 7879–85 (2004).

22. Grundy, A., et al., 'Increased risk of breast cancer associated with long-term shift work in Canada', *Occupational and Environmental Medicine* **70**, 831–8 (2013).

23. The UK's National Health Service discusses the issue here: http://www.nhs.uk/news/2013/07July/Pages/Long-term-night-shifts-can-double-breast-cancer-risk.aspx.

24. Cuesta, M., Boudreau, P., Dubeau-Laramee, G., Cermakian, N., & Boivin, D. B., 'Simulated Night Shift Disrupts Circadian Rhythms of Immune Functions in Humans', *The Journal of Immunology* **196**, 2466–75 (2016).

25. Foster, R. G., et al., 'Circadian photoreception in the retinally degenerate mouse (rd/rd)', *Journal of Comparitive Physiology A* **169**, 39–50 (1991).

26. Russell Foster interviewed by Marie McNeely, for 'People behind the science', 7 March 2016, available online here: http://www.peoplebehindthescience.com/dr-russell-foster/.

27. Freedman, M. S., et al., 'Regulation of mammalian circadian behavior by non-rod, non-cone, ocular photoreceptors', *Science* **284**, 502–4 (1999); Lucas, R. J., Freedman, M. S., Munoz, M.,

Garcia-Fernandez, J. M., & Foster, R. G., 'Regulation of the mammalian pineal by non-rod, non-cone, ocular photoreceptors', *Science* **284**, 505–7 (1999).

28. 'Newswalk: Sleep scientist Russell Foster on how he stopped seeing life in black and white', *Newsweek*, 6 May 2015.

29. O'Neill, J. S., & Reddy, A. B., 'Circadian clocks in human red blood cells', *Nature* **469**, 498–503 (2011).

30. Barger, L. K., et al., 'Prevalence of sleep deficiency and use of hypnotic drugs in astronauts before, during, and after spaceflight: an observational study', *Lancet Neurol* **13**, 904–12 (2014).

31. Crucian, B. E., et al., 'Plasma cytokine concentrations indicate that in vivo hormonal regulation of immunity is altered during long-duration spaceflight', *Journal of Interferon and Cytokine Research* **34**, 778–86 (2014).

32. Crucian, B., et al., 'Alterations in adaptive immunity persist during long-duration spaceflight', *npj Microgravity* **1**, 15013 (2015).

33. Exposure to radiation is predicted to cause a slight increase in an astronaut's lifetime risk of cancer, but this prediction is based on, for example, the occurrence of cancer in survivors of the atomic bomb in Japan, which is not directly comparable.

34. Chang, K., 'Beings not made for space', *New York Times*, 27 January 2014.

35. Interview with Brian Crucian, 24 June 2016.

36. Mehta, S. K., et al., 'Reactivation of latent viruses is associated with increased plasma cytokines in astronauts', *Cytokine* **61**, 205–9 (2013).

37. Crucian, B., et al., 'A case of persistent skin rash and rhinitis with immune system dysregulation onboard the International Space Station', *Journal of Allergy and Clinical Immunology: In Practice* **4**, 759–762 (2016).

38. Ibid.

39. Ibid.

40. Interview with Brian Crucian, 24 June 2016.

41. Ibid.

42. Durrington et al.

43. Wallace, A., Chinn, D., & Rubin, G., 'Taking simvastatin in the morning compared with in the evening: randomised controlled trial', *The British Medical Journal* **327**, 788 (2003).

44. Zhang, R., Lahens, N. F., Ballance, H. I., Hughes, M. E., & Hogenesch, J. B., 'A circadian gene expression atlas in mammals:

implications for biology and medicine', *Proceedings of the National Academy of Sciences USA* **111**, 16219–24 (2014).

45. Brown, M. T., & Bussell, J. K., 'Medication adherence: WHO cares?', *Mayo Clinic Proceedings* **86**, 304–14 (2011).

46. Lin, S., et al., 'Stretchable Hydrogel Electronics and Devices', *Advanced Materials* (2015).

47. In some countries, of course, reaching people in need of a vaccine is still a challenge, never mind vaccinating them at a particular time of day. A blog by Jo Revill, chief executive of the British Society for Immunology, 'Polio vaccination: Real world challenges and solutions', 7 June 2016, highlights the difficulties in delivering the polio vaccine to everyone who needs it in remote locations, available online here: http://britsocimmblog.org/polio-vaccination/.

48. Phillips, A. C., Gallagher, S., Carroll, D., & Drayson, M., 'Preliminary evidence that morning vaccination is associated with an enhanced antibody response in men', *Psychophysiology* **45**, 663–6 (2008).

49. There are many hints that the immune system does behave slightly differently in men and women. For example, some auto-immune diseases are more frequent in women. This may relate to the effects of hormones on the immune system but research to test this is difficult because gender-based differences could also arise from any number of social, economic or cultural factors.

50. Karabay, O., et al., 'Influence of circadian rhythm on the efficacy of the hepatitis B vaccination', *Vaccine* **26**, 1143–4 (2008).

51. Silver, A. C., Arjona, A., Walker, W. E., & Fikrig, E., 'The circadian clock controls toll-like receptor 9–mediated innate and adaptive immunity', *Immunity* **36**, 251–61 (2012).

52. 'Global Health and Aging', a report from the National Institute on Aging (USA) and World Health Organization, available online here: https://www.nia.nih.gov/research/publication/global-health-and-aging/preface.

53. The Office for National Statistics produces annual population data for the UK, available online here: https://www.ons.gov.uk/peoplepopulationandcommunity/populationandmigration/populationestimates. The charity Age UK also produce a monthly collection of statistics about elderly people, available online here: http://www.ageuk.org.uk/professional-resources-home/.

54. Shaw, A. C., Goldstein, D. R., & Montgomery, R. R., 'Age-dependent dysregulation of innate immunity', *Nature Reviews Immunology* **13**, 875–87 (2013).

55. Dorshkind, K., Montecino-Rodriguez, E., & Signer, R. A., 'The ageing immune system: is it ever too old to become young again?', *Nature Reviews Immunology* **9**, 57–62 (2009).

56. Treanor, J. J., et al., 'Effectiveness of seasonal influenza vaccines in the United States during a season with circulation of all three vaccine strains', *Clinical Infectious Diseases: An official publication of the Infectious Diseases Society of America* **55**, 951–9 (2012).

57. By midlife, the telomeres of our DNA are roughly half as short as they were at birth, and by sixty-five, our telomeres have shrunk to be half as short again.

58. Harley, C. B., 'Telomerase and cancer therapeutics', *Nature Reviews Cancer* **8**, 167–79 (2008).

59. Blackburn, E., & Epel, E., *The Telomere Effect: A Revolutionary Approach to Living Younger, Healthier, Longer* (Orion Spring, 2017).

60. Carlson, L. E., et al., 'Mindfulness-based cancer recovery and supportive-expressive therapy maintain telomere length relative to controls in distressed breast cancer survivors', *Cancer* **121**, 476–84 (2015).

61. There are many different subtypes of apoptosis and understanding cell death is an important area of contemporary research.

62. Munoz-Espin, D., & Serrano, M., 'Cellular senescence: from physiology to pathology', *Nature Reviews Molecular Cell Biology* **15**, 482–96 (2014).

63. Baker, D. J., et al., 'Clearance of p16Ink4a-positive senescent cells delays ageing-associated disorders', *Nature* **479**, 232–6 (2011).

64. Kirkwood, T. B., & Austad, S. N., 'Why do we age?', *Nature* **408**, 233–8 (2000).

65. Shaw et al.

66. Discussion with Steve Marsh, 29 April 2016.

67. Sapey, E., et al., 'Phosphoinositide 3–kinase inhibition restores neutrophil accuracy in the elderly: toward targeted treatments for immunosenescence', *Blood* **123**, 239–48 (2014).

68. Shaw et al.

69. Jamieson, B. D., et al., 'Generation of functional thymocytes in the human adult', *Immunity* **10**, 569–75 (1999).

70. Interview with Janet Lord, 23 June 2016.

71. Mark Davis was one of the pioneers who worked out how T cells detect signs of disease in the body and his work on this is discussed in more detail in my first book, *The Compatibility Gene* (Allen Lane, 2013).

72. Brodin, P., et al., 'Variation in the human immune system is largely driven by non-heritable influences', *Cell* **160**, 37–47 (2015).

73. Brodin, P., & Davis, M. M., 'Human immune system variation', *Nature Reviews Immunology* **17**, 21–9 (2017).

74. Furman, D., et al., 'Cytomegalovirus infection enhances the immune response to influenza', *Science Translational Medicine* **7**, 281ra243 (2015).

75. Leng, J., et al., 'Efficacy of a vaccine that links viral epitopes to flagellin in protecting aged mice from influenza viral infection', *Vaccine* **29**, 8147–55 (2011).

76. Taylor, D. N., et al., 'Induction of a potent immune response in the elderly using the TLR-5 agonist, flagellin, with a recombinant hemagglutinin influenza-flagellin fusion vaccine (VAX125, STF2. HA1 SI)', *Vaccine* **29**, 4897–902 (2011).

77. Long, J. E., et al., 'Morning vaccination enhances antibody response over afternoon vaccination: A cluster-randomised trial', *Vaccine* **34**, 2679–85 (2016).

78. Interview with Janet Lord, 23 June 2016.

79. Interview with Akhilesh Reddy, 1 August 2016.

80. Interview with Janet Lord, 23 June 2016.

81. Interview with Arne Akbar, 29 April 2016.

82. Aldrin, B., & Abraham, K., *No Dream is Too High: Life Lessons From a Man Who Walked on the Moon* (National Geographic, 2016).

Chapter Seven

1. 'Autoimmune disease', *Nature Biotechnology* **18 Suppl**, IT7–9 (2000).

2. Davis, D. M., *The Compatibility Gene* (Allen Lane, 2013).

3. Anderson, W., & Mackay, I. R., *Intolerant Bodies: A Short History of Autoimmunity* (Johns Hopkins University Press, 2014).

4. Mackay, I. R., 'Travels and travails of autoimmunity: a historical journey from discovery to rediscovery', *Autoimmunity Reviews* **9**, A251–8 (2010).

5. Aoki, C. A., et al., 'NOD mice and autoimmunity', *Autoimmunity Reviews* **4**, 373–9 (2005).

6. Interview with Shimon Sakaguchi, 14 July 2016.

7. Nishizuka, Y., & Sakakura, T., 'Thymus and reproduction: sex-linked dysgenesis of the gonad after neonatal thymectomy in mice', *Science* **166**, 753–5 (1969).

8. Kojima, A., & Prehn, R. T., 'Genetic susceptibility to post-thymectomy autoimmune diseases in mice', *Immunogenetics* **14**, 15–27 (1981).

9. Interview with Shimon Sakaguchi, 14 July 2016.

10. Sakaguchi, S., Takahashi, T., & Nishizuka, Y., 'Study on cellular events in post-thymectomy autoimmune oophoritis in mice. II. Requirement of Lyt-1 cells in normal female mice for the prevention of oophoritis', *The Journal of Experimental Medicine* **156**, 1577–86 (1982).

11. Germain, R. N., 'Special regulatory T-cell review: A rose by any other name: from suppressor T cells to Tregs, approbation to unbridled enthusiasm', *Immunology* **123**, 20–7 (2008).

12. Benacerraf, B., 'Obituary: Richard Gershon, 1932–1983', *The Journal of Immunology* **131**, 3096–7 (1983).

13. Gershon, R. K., Cohen, P., Hencin, R., & Liebhaber, S. A., 'Suppressor T cells', *The Journal of Immunology* **108**, 586–90 (1972).

14. Interview with Shimon Sakaguchi, 14 July 2016.

15. Benacerraf.

16. Waggoner, W. H., 'Dr Richard Gershon, leader in research on immune system', *New York Times*, 13 July 1983.

17. Germain.

18. Interview with Shimon Sakaguchi, 14 July 2016.

19. Ibid.

20. The famous opening line of *The Go-Between* by L. P. Hartley (Hamish Hamilton, 1953).

21. Kronenberg, M., et al., 'RNA transcripts for I-J polypeptides are apparently not encoded between the I-A and I-E subregions of the murine major histocompatibility complex', *Proceedings of the National Academy of Sciences USA* **80**, 5704–8 (1983).

22. Germain.

23. Green, D. R., & Webb, D. R., 'Saying the "S" word in public', *Immunology Today* **14**, 523–5 (1993).

24. Bloom, B. R., Salgame, P., & Diamond, B., 'Revisiting and revising suppressor T cells', *Immunology Today* **13**, 131–6 (1992).

25. Interview with Fiona Powrie, 16 September 2016.

26. Powrie, F., & Mason, D., 'OX-22high CD4+ T cells induce wasting disease with multiple organ pathology: prevention by the OX-22low subset', *The Journal of Experimental Medicine* **172**, 1701–8 (1990).

27. This was useful to do because mice were studied far more commonly than rats and there were many more tools available for studying the mouse immune system.

28. Powrie, F., Leach, M. W., Mauze, S., Caddle, L. B., & Coffman, R. L., 'Phenotypically distinct subsets of CD4+ T cells induce or protect from chronic intestinal inflammation in C. B-17 scid mice', *International Immunology* **5**, 1461–71 (1993).

29. Morrissey, P. J., Charrier, K., Braddy, S., Liggitt, D., & Watson, J. D., 'CD4+ T cells that express high levels of CD45RB induce wasting disease when transferred into congenic severe combined immunodeficient mice. Disease development is prevented by cotransfer of purified CD4+ T cells', *The Journal of Experimental Medicine* **178**, 237–44 (1993).

30. Sakaguchi, S., Sakaguchi, N., Asano, M., Itoh, M., & Toda, M., 'Immunologic self-tolerance maintained by activated T cells expressing IL-2 receptor alpha-chains (CD25). Breakdown of a single mechanism of self-tolerance causes various autoimmune diseases', *The Journal of Immunology* **155**, 1151–64 (1995).

31. This discovery has become one of the most highly cited papers in *The Journal of Immunology*, published since 1916, as noted here: http://www.jimmunol.org/site/misc/Centennial/MostCitedPub.html.

32. Shevach, E. M., 'Special regulatory T cell review: How I became a T suppressor/regulatory cell maven', *Immunology* **123**, 3–5 (2008).

33. Shevach had long-term funding from the NIH and so was not reliant on each new idea first needing approval by peer review, as would be the case if his work was funded solely through grant applications.

34. Shevach served as the editor-in-chief at *The Journal of Immunology* from 1987 to 1992. He describes why he took on the post, and the challenges of being a scientific editor, in an interview for the American Association of Immunologists recorded on 16 December 2015, available online here: https://vimeo.com/158976383. In this interview, Shevach emphasises how it's important that an editor operates in an impersonal way. As editor, he was not keen to discuss papers with scientists informally over the phone and a

policy of the journal at the time was to not publish its phone number. But of course, it was well known that he was the editor and scientists who had their paper rejected by the journal would sometimes even call his home to complain about it. Later the journal published its phone number prominently, so that aggrieved scientists wouldn't call Shevach or his family at home.

35. Interview with Shimon Sakaguchi, 14 July 2016.

36. Thornton, A. M., & Shevach, E. M., 'CD4+CD25+ immuno-regulatory T cells suppress polyclonal T cell activation in vitro by inhibiting interleukin 2 production', *The Journal of Experimental Medicine* **188**, 287–96 (1998); Takahashi, T., et al., 'Immunologic self-tolerance maintained by CD25+CD4+ naturally anergic and suppressive T cells: induction of autoimmune disease by breaking their anergic/suppressive state', *International Immunology* **10**, 1969–80 (1998).

37. Shevach, E. M., 'The resurrection of T cell-mediated suppression', *The Journal of Immunology* **186**, 3805–7 (2011).

38. Shevach, E. M., 'Certified professionals: CD4(+)CD25(+) suppressor T cells', *The Journal of Experimental Medicine* **193**, F41–6 (2001).

39. Germain.

40. Ibid.

41. Another view, used in some scientific publications, is that suppressor T cells were not just renamed. Rather, they were cells with specific characteristics which were proved to not exist. With this view, regulatory T cells are not the same as suppressor T cells, but rather cells with different traits, which do exist.

42. In 2016, I asked Sakaguchi if he preferred working on an idea that went against the dogma of the time, or did he prefer it once everyone else had caught up and it became a mainstream research area. He replied that it's nice that everyone else now realises the importance of regulatory T cells, but on the other hand, when only a few scientists believed in them, it was very easy to follow everything that was going on. 'Now, I can't follow all the publications.'

43. Russell, L. B., 'The Mouse House: a brief history of the ORNL mouse-genetics program, 1947–2009', *Mutation Research* **753**, 69–90 (2013).

44. Ramsdell, F., & Ziegler, S. F., 'FOXP3 and scurfy: how it all began', *Nature Reviews Immunology* **14**, 343–9 (2014).

45. Godfrey, V. L., Wilkinson, J. E., Rinchik, E. M., & Russell, L. B., 'Fatal lymphoreticular disease in the scurfy (sf) mouse requires T cells that mature in a sf thymic environment: potential model for thymic education', *Proceedings of the National Academy of Sciences USA* **88**, 5528–32 (1991).

46. Brunkow, M. E., et al., 'Disruption of a new forkhead/winged-helix protein, scurfin, results in the fatal lymphoproliferative disorder of the scurfy mouse', *Nature Genetics* **27**, 68–73 (2001).

47. Ramsdell & Ziegler.

48. Bennett, C. L., et al., 'The immune dysregulation, polyendocrinopathy, enteropathy, X-linked syndrome (IPEX) is caused by mutations of FOXP3', *Nature Genetics* **27**, 20–1 (2001).

49. Sakaguchi, S., Wing, K., & Miyara, M., 'Regulatory T cells – a brief history and perspective', *European Journal of Immunology* **37** **Suppl 1**, S116–23 (2007).

50. These three scientists together were rewarded for their work with the prestigious 2017 Crafoord Prize, awarded by the the Royal Swedish Academy of Sciences, for which the prize money is 6 million Swedish krona (approximately £550,000): http://www.crafoordprize.se/press/arkivpressreleases/thecrafoordprizeinpolyarthritis2017.5.470b0073156f7766c064a8.html.

51. Hori, S., Nomura, T., & Sakaguchi, S., 'Control of regulatory T cell development by the transcription factor Foxp3', *Science* **299**, 1057–61 (2003); Fontenot, J. D., Gavin, M. A., & Rudensky, A. Y., 'Foxp3 programs the development and function of CD4+CD25+ regulatory T cells', *Nature Immunology* **4**, 330–6 (2003); Khattri, R., Cox, T., Yasayko, S. A., & Ramsdell, F., 'An essential role for Scurfin in CD4+CD25+ T regulatory cells', *Nature Immunology* **4**, 337–42 (2003).

52. Ramsdell & Ziegler.

53. Fiona Powrie's lecture at the Academy of Medical Sciences, London, 2 December 2014, in memory of Jean Shanks. Available online here: https://www.youtube.com/watch?v=rvEdEw0CU80.

54. Sender, R., Fuchs, S., & Milo, R., 'Revised Estimates for the Number of Human and Bacteria Cells in the Body', *PLoS Biology* **14**, e1002533 (2016).

55. Zeevi, D., Korem, T., & Segal, E., 'Talking about cross-talk: the immune system and the microbiome', *Genome Biology* **17**, 50 (2016).

56. Arpaia, N., & Rudensky, A. Y., 'Microbial metabolites control gut inflammatory responses', *Proceedings of the National Academy of Sciences USA* **111**, 2058–9 (2014).

57. Chang, P. V., Hao, L., Offermanns, S., & Medzhitov, R., 'The microbial metabolite butyrate regulates intestinal macrophage function via histone deacetylase inhibition', *Proceedings of the National Academy of Sciences USA* **111**, 2247–52 (2014).

58. Chan, J. K., et al., 'Alarmins: awaiting a clinical response', *Journal of Clinical Investigation* **122**, 2711–19 (2012).

59. A BBC documentary about Polly Matzinger, *Turned On By Danger*, was broadcast in 1997, in the Horizon series.

60. Matzinger, P., 'Tolerance, danger, and the extended family', *Annual Review of Immunology* **12**, 991–1045 (1994).

61. Discussion with Polly Matzinger, 14 December 2011.

62. Silverstein, A. M., 'Immunological tolerance', *Science* **272**, 1405–8 (1996).

63. Cooper, G., 'Clever bunny', *Independent*, 17 April 1997.

64. Matzinger, P., & Mirkwood, G., 'In a fully H-2 incompatible chimera, T cells of donor origin can respond to minor histocompatibility antigens in association with either donor or host H-2 type', *The Journal of Experimental Medicine* **148**, 84–92 (1978).

65. Vance, R. E., 'Cutting edge commentary: a Copernican revolution? Doubts about the danger theory', *The Journal of Immunology* **165**, 1725–8 (2000).

66. Schiering, C., et al., 'The alarmin IL-33 promotes regulatory T-cell function in the intestine', *Nature* **513**, 564–8 (2014).

67. Martin, N. T., & Martin, M. U., 'Interleukin 33 is a guardian of barriers and a local alarmin', *Nature Immunology* **17**, 122–31 (2016).

68. Aune, D., et al., 'Dietary fibre, whole grains, and risk of colorectal cancer: systematic review and dose-response meta-analysis of prospective studies', *The British Medical Journal* **343**, d6617 (2011).

69. Bollrath, J., & Powrie, F., 'Feed your Tregs more fiber', *Science* **341**, 463–4 (2013).

70. Furusawa, Y., et al., 'Commensal microbe-derived butyrate induces the differentiation of colonic regulatory T cells', *Nature* **504**, 446–50 (2013); Arpaia, N., et al., 'Metabolites produced by commensal bacteria promote peripheral regulatory T-cell generation', *Nature* **504**, 451–5 (2013).

71. Ohnmacht, C., et al., 'The microbiota regulates type 2 immunity through RORgammat(+) T cells', *Science* **349**, 989–93 (2015).

72. Strachan, D. P., 'Hay fever, hygiene, and household size', *The British Medical Journal* **299**, 1259–60 (1989).

73. Chatila, T. A., 'Innate Immunity in Asthma', *New England Journal of Medicine* **375**, 477–9 (2016).

74. Stein, M. M., et al., 'Innate Immunity and Asthma Risk in Amish and Hutterite Farm Children', *New England Journal of Medicine* **375**, 411–21 (2016).

75. Ibid.

76. Tanner, L., 'Can house dust explain why Amish protected from asthma?', *Washington Post*, 3 August 2016.

77. Blaser, M., *Missing Microbes: How Killing Bacteria Creates Modern Plagues* (Oneworld Publications, 2014).

78. Korpela, K., et al., 'Intestinal microbiome is related to lifetime antibiotic use in Finnish pre-school children', *Nature Communications* **7**, 10410 (2016).

79. Ortqvist, A. K., et al., 'Antibiotics in fetal and early life and subsequent childhood asthma: nationwide population based study with sibling analysis', *The British Medical Journal* **349**, g6979 (2014).

80. Vatanen, T., et al., 'Variation in Microbiome LPS Immunogenicity Contributes to Autoimmunity in Humans', *Cell* **165**, 842–53 (2016).

81. Hofer, U., 'Microbiome: Is LPS the key to the hygiene hypothesis?', *Nature Reviews Microbiology* **14**, 334–5 (2016).

82. Bollrath & Powrie.

83. As Gabriel Núñez from the University of Michigan puts it: 'The use of probiotics in the clinic has been often associated with controversial or negative results. In my view, these disappointing results reflect the fact that the choice of probiotics has largely relied on empirical results with little or no scientific rationale for the selection of specific bacterial species or strains.' In Underhill, D. M., Gordon, S., Imhof, B. A., Núnez, G., & Bousso, P., 'Elie Metchnikoff (1845–1916): celebrating 100 years of cellular immunology and beyond', *Nature Reviews Immunology* (2016).

84. Steidler, L., et al., 'Treatment of murine colitis by Lactococcus lactis secreting interleukin-10', *Science* **289**, 1352–5 (2000).

85. Horowitz, A., et al., 'Genetic and environmental determinants of human NK cell diversity revealed by mass cytometry', *Science Translational Medicine* **5**, 208ra145 (2013).

Chapter Eight

1. Grady, D., 'Harnessing the immune system to fight cancer', *New York Times*, 30 July 2016.
2. Sharma, P., & Allison, J. P., 'The future of immune checkpoint therapy', *Science* **348**, 56–61 (2015).
3. Sharon Belvin, speaking in 'Advancing the next wave of cancer therapy', a video from the Cancer Research Insitute, New York. Available online here: http://www.cancerresearch.org/news-publications/video-gallery/advancing-the-next-wave-of-cancer-therapy.
4. 'A Scientist's Dream Fulfilled: Harnessing the Immune System to Fight Cancer', National Public Radio, USA, 9 June 2016. Available online here: http://www.npr.org/sections/health-shots/2016/06/09/480435066/a-scientists-dream-fulfilled-harnessing-the-immune-system-to-fight-cancer.
5. Her physician, Jedd Wolchok, asked Allison to come and visit her in his clinic.
6. Video made to celebrate and describe the research which received the 2015 Lasker DeBakey Clinical Medical Research Award, published on 7 September 2015. Available online here: https://www.youtube.com/watch?v=W8fUAvENkCo&feature=youtu.be.
7. Gross, L., 'Intradermal Immunization of C3H Mice against a Sarcoma That Originated in an Animal of the Same Line', *Cancer Research* **3**, 326–33 (1943).
8. Earlier than this, in the 1930s and early 1940s, Peter Gorer, George Snell and others, showed that tumours taken from one mouse would be killed when implanted in a different (unrelated) mouse, but their framework for thinking about this was in terms of transplant rejection, rather than a specific immune response against cancer.
9. Shankaran, V., et al., 'IFNgamma and lymphocytes prevent primary tumour development and shape tumour immunogenicity', *Nature* **410**, 1107–11 (2001).
10. Coulie, P. G., Van den Eynde, B. J., van der Bruggen, P., & Boon, T., 'Tumour antigens recognized by T lymphocytes: at the core of cancer immunotherapy', *Nature Reviews Cancer* **14**, 135–46 (2014).
11. Radiation therapy was introduced soon after, in 1896, which was more easily administered and offered more consistent results. This

was probably one of the reasons that Coley's toxins were not used or studied as widely as they might have been at another time.

12. Engelking, C., 'Germ of an idea: William Coley's cancer-killing toxins', *Discover Magazine*, April 2016.

13. Cancer Research UK, 'What is Coley's toxins treatment for cancer?' Available online here: http://www.cancerresearchuk. org/about-cancer/cancers-in-general/cancer-questions/coleys-toxins-cancer-treatment.

14. 'Science Webinar: Targeting Cancer Pathways, Part 5: Understanding Immune Checkpoints', 19 January 2016. Available online: http://webinar.sciencemag.org/webinar/archive/part-5-targeting-cancer-pathways.

15. Video made to celebrate and describe the research which received the 2015 Lasker DeBakey Clinical Medical Research Award, published on 7 September 2015. Available online here: https://www.youtube.com/watch?v=W8fUAvENkCo&feature=youtu.be.

16. 'The *Journal of Clinical Investigations*' Conversations with Giants in Medicine: James Allison', 4 January 2016. Available online here: https://www.youtube.com/watch?v=yCi0bUDR7KA.

17. A line attributed to the Chinese philosopher Lao Tzu.

18. Without this second signal from co-stimulatory proteins to signify the presence of germs, T cells that receive a signal through their T cell receptor don't just not react, they become desensitised, or anergic, so that they are unable to participate in an immune response. This helps safeguard T cells from reacting to healthy cells and tissues. Ronald Schwartz and Marc Jenkins at the National Institutes of Health in the USA, as well as many others, established this.

19. Grady.

20. Brunet, J. F., et al., 'A new member of the immunoglobulin super-family – CTLA-4', *Nature* **328**, 267–70 (1987).

21. Bluestone, J. A., 'CTLA-4Ig is finally making it: a personal perspective', *American Journal of Transplantation* **5**, 423–4 (2005).

22. Prasad, V., 'The Folly of Big Science Awards', *New York Times*, 3 October 2015.

23. Littman, D. R., 'Releasing the Brakes on Cancer Immunotherapy', *Cell* **162**, 1186–90 (2015).

24. Price, P., 'Tested: A reboot for the immune system', *Popular Science*, 15 March 2010.

25. Walunas, T. L., et al., 'CTLA-4 can function as a negative regulator of T cell activation', *Immunity* **1**, 405–13 (1994).

26. Interview with Jeffrey Bluestone, 23 November 2016.

27. Laurie Glimcher and Abul Abbas.

28. Interview with Jeffrey Bluestone, 23 November 2016.

29. Discussion with Matthew 'Max' Krummel, 21 September 2016.

30. Interview with Matthew 'Max' Krummel, 28 October 2016.

31. Krummel, M. F., & Allison, J. P., 'CD28 and CTLA-4 have opposing effects on the response of T cells to stimulation', *The Journal of Experimental Medicine* **182**, 459–65 (1995).

32. A technical detail here is that Bluestone's and Allison's teams tried to resolve this issue by comparing what happened when they used different forms of the antibody, including fragments of it, so-called Fab fragments, which are unlikely to trigger the receptor and more likely to block it.

33. Tivol, E. A., et al., 'Loss of CTLA-4 leads to massive lympho-proliferation and fatal multiorgan tissue destruction, revealing a critical negative regulatory role of CTLA-4', *Immunity* **3**, 541–7 (1995); Waterhouse, P., et al., 'Lymphoproliferative disorders with early lethality in mice deficient in Ctla-4', *Science* **270**, 985–8 (1995).

34. Krummel, M. F., Sullivan, T. J., & Allison, J. P., 'Superantigen responses and co-stimulation: CD28 and CTLA-4 have opposing effects on T cell expansion in vitro and in vivo', *International Immunology* **8**, 519–23 (1996).

35. Allison, J. P., 'Checkpoints', *Cell* **162**, 1202–5 (2015).

36. Leach, D. R., Krummel, M. F., & Allison, J. P., 'Enhancement of antitumor immunity by CTLA-4 blockade', *Science* **271**, 1734–6 (1996).

37. 'The *Journal of Clinical Investigations*' Conversations with Giants in Medicine: James Allison', 4 January 2016. Available online here: https://www.youtube.com/watch?v=yCi0bUDR7KA.

38. Allison.

39. 'The 2013 Novartis Prize for Clinical Immunology', *Cancer Immunology Research* **1**, 285–7 (2013).

40. Video made to celebrate and describe the research which received the 2015 Lasker DeBakey Clinical Medical Research Award, published on 7 September 2015. Available online here: https://www.youtube.com/watch?v=W8fUAvENkCo&feature=youtu.be.

41. 'The *Journal of Clinical Investigations*' Conversations with Giants in Medicine: James Allison', 4 January 2016. Available online here: https://www.youtube.com/watch?v=yCi0bUDR7KA.

42. Littman.

43. Ibid.

44. Hoos, A., 'Development of immuno-oncology drugs – from CTLA4 to PD1 to the next generations', *Nature Reviews Drug Discovery* **15**, 235–47 (2016).

45. Wolchok, J. D., et al., 'Guidelines for the evaluation of immune therapy activity in solid tumors: immune-related response criteria', *Clinical Cancer Research* **15**, 7412–20 (2009).

46. Hoos.

47. Littman.

48. Hoos.

49. 'OncoImmune Announces Option and License Agreement with Pfizer Inc.', company announcement, 15 September 2016. Available online here: http://announce.ft.com/detail?dockey =600–201609150900BIZWIRE_USPRX____BW5151–1.

50. Morse, A., 'Bristol to Acquire Medarex', *Wall Street Journal*, 23 July 2009.

51. Hodi, F. S., et al., 'Improved survival with ipilimumab in patients with metastatic melanoma', *New England Journal of Medicine* **363**, 711–23 (2010).

52. Schadendorf, D., et al., 'Pooled Analysis of Long-Term Survival Data From Phase II and Phase III Trials of Ipilimumab in Unresectable or Metastatic Melanoma', *Journal of Clinical Oncology* **33**, 1889–94 (2015).

53. Sondak, V. K., Smalley, K. S., Kudchadkar, R., Grippon, S., & Kirkpatrick, P., 'Ipilimumab', *Nature Reviews Drug Discovery* **10**, 411–12 (2011).

54. Sales of Yervoy in 2015 reported by Bristol-Myers Squibb are available online here: https://www.bms.com/ourcompany/ Pages/keyfacts.aspx.

55. Hoos.

56. Video made to celebrate and describe the research which received the 2015 Lasker DeBakey Clinical Medical Research Award, published on 7 September 2015. Available online here: https://www.youtube.com/watch?v=W8fUAvENkCo&feature =youtu.be.

57. Ishida, Y., Agata, Y., Shibahara, K., & Honjo, T., 'Induced expression of PD-1, a novel member of the immunoglobulin gene superfamily, upon programmed cell death', *EMBO Journal* **11**, 3887–95 (1992).

58. Nishimura, H., Nose, M., Hiai, H., Minato, N., & Honjo, T., 'Development of lupus-like autoimmune diseases by disruption of the PD-1 gene encoding an ITIM motif-carrying immunoreceptor', *Immunity* **11**, 141–51 (1999).

59. Okazaki, T., & Honjo, T., 'PD-1 and PD-1 ligands: from discovery to clinical application', *International Immunology* **19**, 813–24 (2007).

60. Hoos.

61. Robert, C., et al., 'Pembrolizumab versus Ipilimumab in Advanced Melanoma', *New England Journal of Medicine* **372**, 2521–32 (2015).

62. Ansell, S. M., et al., 'PD-1 blockade with nivolumab in relapsed or refractory Hodgkin's lymphoma', *New England Journal of Medicine* **372**, 311–19 (2015).

63. Long, E. O., 'Negative signaling by inhibitory receptors: the NK cell paradigm', *Immunolical Reviews* **224**, 70–84 (2008).

64. Interview with Eric Vivier, 4 October 2016.

65. Meng, X., Huang, Z., Teng, F., Xing, L., & Yu, J., 'Predictive biomarkers in PD-1/PD-L1 checkpoint blockade immunotherapy', *Cancer Treatment Reviews* **41**, 868–76 (2015).

66. Qureshi, O. S., et al., 'Trans-endocytosis of CD80 and CD86: a molecular basis for the cell-extrinsic function of CTLA-4', *Science* **332**, 600–3 (2011).

67. Schneider, H., et al., 'Reversal of the TCR stop signal by CTLA-4', *Science* **313**, 1972–5 (2006).

68. Davis, D. M., 'Mechanisms and functions for the duration of intercellular contacts made by lymphocytes', *Nature Reviews Immunology* **9**, 543–55 (2009).

69. Email correspondence with Christopher Rudd, 25 October 2016.

70. Schneider, H., & Rudd, C. E., 'Diverse mechanisms regulate the surface expression of immunotherapeutic target ctla-4', *Frontiers in Immunology* **5**, 619 (2014).

71. Moynihan, K. D., et al., 'Eradication of large established tumors in mice by combination immunotherapy that engages innate and adaptive immune responses', *Nature Medicine* (2016).

72. Sean Parker is interviewed in 'NBC Dateline On Assignment: Hacking Cancer', 22 May 2016. Available online here: http://www.nbcnews.com/feature/on-assignment/hacking-cancer-n575756.

73. Tom Hanks interviewed for 1WMN TV in 'Sean Parker and the Parker Foundation Launch the Parker Institute For Cancer Immunotherapy', 14 April 2016. Available online here: https://www.youtube.com/watch?v=guVIGDc4z6o.

74. Cha, A. E., 'Sean Parker, Silicon Valley's bad boy genius, wants to kick the *!$% out of cancer', *Washington Post*, 15 April 2016.

75. Leaf, C., 'Can Sean Parker hack cancer?', *Fortune* magazine, 22 April 2016. Available online here: http://fortune.com/digital-health-sean-parker-cancer/.

76. Parker, S., 'Sean Parker: Philanthropy for Hackers', *Wall Street Journal*, 26 June 2015.

77. Interview with Lewis Lanier, 8 November 2016.

78. Interview with Jeffrey Bluestone, 23 November 2016.

79. Interview with Lewis Lanier, 8 November 2016.

80. Jeff Bluestone said this during a speech at Dreamtalk, 'Using Yourself to Beat Cancer and How We Will Beat Zika', 16 October 2016. Available online here: https://www.youtube.com/watch?v=eXAcSloGVGA.

81. Rosenberg, S. A., & Restifo, N. P., 'Adoptive cell transfer as personalized immunotherapy for human cancer', *Science* **348**, 62–8 (2015).

82. Porter, D. L., Levine, B. L., Kalos, M., Bagg, A., & June, C. H., 'Chimeric antigen receptor-modified T cells in chronic lymphoid leukemia', *New England Journal of Medicine* **365**, 725–33 (2011).

83. Zelig Eshar first published the use of T cells genetically modified to contain an antibody-like receptor in 1989, and this was soon followed by work from Leroy Hood and colleagues, published in 1990. Eshar, from the Weizmann Institute in Israel, continued to develop CAR T cells and was influenced by a year-long sabbatical he took in 1990 to work with Steven Rosenberg at the National Institutes of Health, USA.

84. Gill, S., & June, C. H., 'Going viral: chimeric antigen receptor T-cell therapy for hematological malignancies', *Immunological Reviews* **263**, 68–89 (2015).

85. Anonymous, 'Penn Medicine Patient Perspective: I was sure the war was on. I was sure CLL cells were dying', 10 August 2011. Available online here: http://www.uphs.upenn.edu/news/News_Releases/2011/08/t-cells/perspective.html.

86. Ellebrecht, C. T., et al., 'Reengineering chimeric antigen receptor T cells for targeted therapy of autoimmune disease', *Science* **353**, 179–84 (2016).

87. Using, for example, the revolutionary genome-editing tool known as CRISPR–Cas9. Doudna, J., & Sternberg, S., *A Crack in Creation: The New Power to Control Evolution* (The Bodley Head, 2017).

88. Brynner, R., & Stephens, T., *Dark Remedy: The Impact of Thalidomide and its Rival as a Vital Medicine* (Basic Books, 2001).

89. Soon after, in the 1970s, other drugs proved to be more effective. The World Health Organization does not recommend the use of thalidomide for leprosy but sadly, some thalidomide babies are still born each year due to inappropriate use of the drug. Details are here: http://www.who.int/lep/research/thalidomide/en/.

90. Zeldis, J. B., Knight, R., Hussein, M., Chopra, R., & Muller, G., 'A review of the history, properties, and use of the immuno-modulatory compound lenalidomide', *Annals of the New York Academy of Sciences* **1222**, 76–82 (2011).

91. Lagrue, K., Carisey, A., Morgan, D. J., Chopra, R., & Davis, D. M., 'Lenalidomide augments actin remodeling and lowers NK-cell activation thresholds', *Blood* **126**, 50–60 (2015).

92. Many lessons can be learnt from how the world reacted to the 2013–16 outbreak of Ebola. Several articles and books discuss this, such as Evans, N. G., Smith, T. C., & Majumder, M. S. (eds), *Ebola's Message* (MIT Press, 2016).

Index

Abbott Laboratories (pharmaceutical company), 99
accessory cells, 33, 35, 41, 56
acetylcholine, 115
Act of Creation, The (Koestler), 32
Acta Virologica journal, 91–2
adaptive immunity, 17, 30, 45
adjuvants, 10, 14–15, 29–30, 50, 51, 138
adrenal glands, 116–18
ageing, 4, 139–48
aggression, 123
AIDS (acquired immune deficiency syndrome), 123
Akira, Shizuo, 28
alarmins, 162, 164
Albert Lasker Award for Clinical Medical Research, 100
allergies, 135–6, 166–7
Allison, Jim, 172–3, 175–81, 184, 188–9, 191
aluminium salts, 14, 29–30, 50
American Cancer Society, 69
American Society of Clinical Oncology, 71
American Society of Immunology, 20
Amgen (biotechnology company), 87
Amish, 166–7
Andrewes, Christopher, 62
ankylosing spondylitis, 99
anti-cytokines, 89–90, 92–3
anti-TNF therapy, 92–102
antibiotics, 167–8

antibodies, 89–90, 93–9, 102–5, 175, 188–9
antihistamines, 136
Apollo missions, 135
'Approaching the asymptote? Evolution and revolution in immunology' (Janeway), 15, 19–20
aspirin, 114
asthma, 116, 119–20, 131, 137, 167–8
AstraZeneca (pharmaceutical company), 182
Atwood, Margaret, 72
Auron, Philip, 85
autoimmune disease, 30, 32, 78, 139, 149–61, 165, 168–9, 185, 188, 194
    coeliac disease, 151
    multiple sclerosis, 32, 89, 149
    rheumatoid arthritis, 87–9, 95–7, 99–100, 103, 114, 116–17, 119, 132, 175
    type I diabetes, 32, 89, 123, 149, 151

B cells, 16–17, 29, 33, 45, 89–90, 92–3, 102–4
bacteria, 2, 18, 19, 24, 70, 161–2, 164, 167–9
    genetically modified bacteria, 169
Barry, Dan, 64
Belvin, Sharon, 172–3, 183
Berkeley Cancer Research Laboratory, University of California 179
Beutler, Bruce, 24–7, 30, 93–4, 99

Bill and Melinda Gates Foundation, 30
Biogen (biotechnology company), 71
biomarkers, 187–8
Blaine, David, 190
Blobel, Günter, 36
Bluestone, Jeff, 178–9, 191–2, 194
body clock, 131, 133–4
body's daily rhythm, 129–34, 137–8
Boon, Thierry, 173–4
Bottomly, Kim, 10, 14
Brennan, Fionula, 94–5
Bristol-Myers Squibb (pharmaceutical
    company), 183
Burke, Derek, 65
Burnet, Macfarlane, 57, 150

Canada Gairdner International
    Award, 100
cancer, 52–5, 79–82, 89, 104–5,
    110–12, 140–41, 172–7, 180–95
  and interferon, 66, 68–9, 71–2, 76
  leukaemia, 72, 76, 103, 193
  melanoma, 76, 80, 82, 172, 183, 185
  myeloma, 69, 93, 195
  non-Hodgkin lymphoma, 103–4
cancer cells, 109–12, 174
cancer immunotherapy, 82, 177–84
Cancer Research UK, 175
Cantell, Kari, 67–71
CAR (chimeric antigen receptor) T
    cell therapy, 193–4
Caroline of Ansbach, queen of Great
    Britain and Ireland, 12
Celgene (biotechnology company), 195
Cell journal, 23
Centocor (biotechnology company),
    93–9, 102
centrifuges, 37–8
Cerami, Anthony, 93–4
Chabris, Christopher, 34
Chain, Ernst, 65
Charing Cross Hospital, London, 96
checkpoint inhibitors, 186–7, 189,
    192, 194
chemotherapy, 182
chess, 9

chicken-pox, 135
China, 75–6
cholesterol, 137
circadian rhythm, 129–34
Cistron (biotechnology company), 86
Claude, Albert, 36
co-stimulatory proteins, 48, 177, 188
coeliac disease, 151
Cohn, Zanvil, 33, 36
Cold Spring Harbor, New York, 15
Coley, William, 174–5, 181
'Coley's toxins', 174–5
colitis, 99
common cold, 3, 101
Commotion in the Blood, A (Hall), 79
Compatibility Gene, The (Davis), 4, 77
Concordia research station,
    Antarctica, 136
Corden, James, 190
cortisol, 118–19, 121–3, 127, 130, 132
cortisone, 116–19
cowpox, 12
Crick, Francis, 65
Crick Institute, London, 146
Crohn's disease, 78, 99
Crucian, Brian, 136
CTLA-4 (cytotoxic T-lymphocyte-
    associated molecule 4), 177–83,
    185–6, 188–9
cytokines, 73, 77–8, 80–82, 84–7, 89,
    92–100, 101, 105, 114–15, 132,
    161, 175
  anti-cytokines, 89–90, 92–3
  IL-1, 77, 85–7
  IL-2, 77–8, 80–82
  IL-6, 100
  IL-10, 78
cytomegalovirus, 135, 145

Dahl, Roald, 21
Darwin, Charles, 58, 133
Darwin, Erasmus, 58
Davis, Mark, 144–5
dendritic cells, 33–45, 46–53, 55–6,
    74, 105, 153, 155, 177
depression, 127

dexamethasone, 119
diabetes, 32, 89, 123, 149, 151
Dinarello, Charles, 85
diphtheria, 13–14
DNA (deoxyribonucleic acid), 66, 71,
    140, 142, 160
Duke, Charlie, 135
Dutton, Richard, 20–21
de Duve, Christian, 37–8

Ebola virus, 195
Edwin Smith papyrus, 110
Ehrlich, Paul, 89
electron microscopes, 36–7, 60
endocrinology, 152
Enbrel (drug), 99
epidemics, 11, 58
epigenetics, 140
epilepsy, 132
Epstein–Barr virus, 135
Essay on Science and Narcissism, An
    (Lemaitre), 27
Estonia, 168
eyes, 133–4

Feldmann, Sir Marc, 83–4, 87–90,
    94–100, 101–2, 104
fever, 112–15
Feynman, Richard, 1
Finland, 168
flaggelin, 145
Flash Gordon, 64
Fleming, Alexander, 64
flu *see* influenza
Food and Drug Administration
    (FDA), 52, 53, 55
Foster, Russell, 133–4
Foxp3 gene, 160–61, 163
Fred Hutchinson Cancer Research
    Center, Seattle, 85
fruit flies, 21–3, 160
Fullam, Ernest, 36

Galileo Galilei 35
Genentech (biotechnology
    company), 52, 70, 92

genes, 22–5, 47–8, 70–71, 84–6, 91,
    131, 140–42, 160–61
    epigenetics, 140
    Foxp3 gene, 160–61, 163
    IFITM3 gene, 75–6
    IL-1 gene, 85–7
    interferon-stimulated genes, 74–6
    major histocompatibility complex
        (MHC) genes, 47
    TLR4 gene, 24–5, 27–8
    toll genes, 22–4, 27
    toll-like receptor (TLR) genes,
        23–5, 27–8, 30, 42
genetic analysis, 76
genetic engineering, 70
genetic testing, 188
GenPharm (pharmaceutical
    company), 181
George I, king of Great Britain and
    Ireland, 11
George II, king of Great Britain and
    Ireland, 12
Germain, Ron, 159
Gershon, Richard, 153–4
Gillis, Steven, 85–7
Glenny, Alexander, 13–14
Golstein, Pierre, 177
Gordon, Susanna, 64
gout, 131–2
Gresser, Ion, 66–7
Grünenthal (pharmaceutical
    company), 194
Gutterman, Jordan, 69

Hall, Stephen S., 79
Hanks, Tom, 190
Hartley, L. P., 155
Harvard University, 9, 34, 109, 199
Hawn, Goldie, 190
hay fever, 116, 165–6
Hench, Philip, 115–20
Henney, Christopher, 85–7
hepatitis A virus, 137
hepatitis B virus, 73, 77, 102, 138
hepatitis C virus, 77
herpes virus, 123, 135

HIV (human immunodeficiency virus), 48, 50–51, 74, 123, 127, 186
Hoffmann, Danièle, 22
Hoffmann, Jules, 21–4, 26–7, 29
Honjo, Tasuku, 184
hormones, 73, 111, 114–16, 118, 122, 151–2
human insulin, 70
Hunter, John, 12
Hutterites, 166–7
Huxley, Thomas Henry, 133
hybridomas, 93, 102
hydrocortisone, 119
hygiene hypothesis, 165–7
hyperthermia therapy, 111
hypothalamus, 114–15, 133–4

IFITM3 gene, 75–6
IL-1 gene, 85–7
immune cells, 83–4, 87, 111, 113–15, 142–5, 149–50, 152–4, 160, 162, 169–70, 175, 186, 189, 192
immune checkpoint therapy, 176, 186–7
immune reactions, 15–16, 33, 41–2, 47, 49–53, 78, 94–5, 102, 118, 143, 153–4, 170, 177
Immune-Related Response Criteria, 182–3
Immunex (biotechnology company), 85–6, 99, 156
Immunity journal, 179
Immunology journal, 153
immunotherapy, 82, 174, 190–92
Imperial College London, 76, 133, 134
Inaba, Kayo, 50–51
'inflamm-ageing', 142
influenza virus, 4, 58–62, 74–6, 143, 145–6
innate immunity, 17, 20, 28–30, 45, 74
inoculation, 11–12, 16
insect toll gene, 23–4, 27
insects, 21–4, 120, 131, 160
Institute of Inflammation and Ageing, University of Birmingham, 146

insulin, 36, 151, 169
    human insulin, 70
interferon, 61–73, 74–8, 91
    alpha, 76–7
    gamma, 77
    interferon-stimulated genes, 74–5
interleukins, 77–8
    see also cytokines
IPEX (immune dysregulation, polyendocrinopathy, enteropathy, X-linked syndrome), 160–61
ipilimumab, 183
Isaacs, Alick, 57–66, 72–3, 91–2

Janeway, Charles, 9–10, 13–15, 17–21, 23, 25, 26–7, 30–31, 42, 54, 163, 177
Japan, 75–6
Japanese Cancer Research Institute, 72
jaundice, 116
Jenner, Edward, 12–13, 29
jet lag, 132
Johnson & Johnson (pharmaceutical company), 98
Joly, Pierre, 21
Journal of Experimental Medicine, 36
Journal of Immunology, 158
June, Carl, 192–4

Kabat-Zinn, Jon, 126
Kendall, Edward, 116–18, 120–21
Kennedy, John F., 147
Koch, Robert, 150
Koestler, Arthur, 32
Köhler, Georges, 92–3
Kondo, Kazunari, 153
Korman, Alan, 181
Krim, Mathilde, 68
Krummel, Matthew 'Max', 179–80
Kyoto University, Japan, 50–51

Lady Gaga, 190
Lamott, Anne, 46
Lancet journal, 87

Langerhans, Paul, 34
Lasker, Mary, 68, 69
laughter, 123–4
Le, Junming 'Jimmy', 93
Leach, Dana, 180
Lemaitre, Bruno, 22–3, 27
leprosy, 195
leukaemia, 72, 76, 103, 193
leukocytes, *see* white blood cells
life expectancy, 139
Life magazine, 69
Lindenmann, Jean, 57–64, 72–3
lipopolysaccharide (LPS), 15, 24–5,
    28, 145
Lord, Janet, 146–7
Lower, William, 35
lymph nodes, 43–4

McGill University, Montreal, 121
macrophages, 40–41, 44–5, 49, 113,
    170, 186
Maini, Sir Ravinder 'Tiny', 88–90,
    96–7, 99–100, 101–2
major histocompatibility complex
    (MHC) genes, 47
Mallory, George, 147
Malo, Danielle, 25–6
Mammalian Genetics Laboratory,
    Oak Ridge, 159
Manhattan Project, 159
Mason, Don, 156
Massachusetts General Hospital,
    Boston, 33
Massachusetts Institute of
    Technology (MIT), 85–7
Mattingly, Ken, 135
Matzinger, Polly, 162–3, 177
Mayo Clinic, Rochester, Minnesota,
    115–18
MD Anderson Cancer Center,
    Houston, Texas, 172
MDX-010, 181, 183
measles, 58, 135
Medarex (biotechnology company),
    181, 183
Medical Research Council, UK, 64

medicines, 137, 138
MedImmune (pharmaceutical
    company), 182
Medzhitov, Ruslan, 19–21, 23–4,
    26–7, 30–31
melanoma, 76, 80, 82, 172, 183, 185
Mellman, Ira, 52–3, 55
Memorial Hospital, New York, 174
Merck (pharmaceutical company),
    117, 118
metabolites, 162, 164
Metchnikoff, Ilya, 39–40
methotrexate, 98
mice, 159–60
    germ-free mice, 165
microbiome, 162–3, 165, 167–9
migraines, 132
Milstein, César, 92–3
mindfulness, 3, 124, 126–7, 140
Moberg, Carol, 36
molecular patterns, 17
monoclonal antibodies, 93
Mooser, Hermann, 63
Moscow State University, 19–20
mosquitoes, 120, 131
multiple sclerosis, 32, 89, 149
myeloma, 69, 93, 195

NASA, 125–6, 135–7, 147
National Cancer Institute, USA,
    78–9
National Health Service (NHS), UK,
    125
National Institute of Medical
    Research, London, 57–8, 62, 64
National Institutes of Health (NIH),
    USA, 26, 84, 125, 158, 159, 163
Natural Killer cells, 77–8, 103–4,
    109–10, 112, 170, 174, 186, 195
Nature journal, 26, 86, 91
Nelmes, Sarah, 12
neutrophils, 77
New England Journal of Medicine, 53,
    183
New Scientist journal, 65
New York Times, 81, 154, 173, 178

New York University Medical School, 90–92
Nexstar (biotechnology company), 181
night-shift work, 3, 132, 134
Nishizuka, Yasuaki, 151–2
Nixon, Richard, 68
Nobel Prizes, 30, 38, 39, 92, 109, 121
  in Physiology or Medicine, 1950, 118, 120
  in Physiology or Medicine, 2011, 26–7, 30, 54–5
non-Hodgkin lymphoma, 103–4
noradrenaline, 114
Nussenzweig, Michel, 41, 52

Oak Ridge National Laboratory, USA, 159–60
Obama, Barack, 100
Openshaw, Peter, 76
organelles, 38
Orthoclone (antibody), 102
Osaka University, Japan, 28

Palade, George, 36–8
papilloma virus, 30
parasitic worms, 44–5
Parker, Sean, 190–91
Parker Institute, USA, 190–92, 194
Pasteur, Louis, 29, 150
pattern-recognition receptors, 17, 21, 28–30, 42–3, 45, 47–9, 73–4, 114
  RIG-1, 28
Paul, Bill, 26
PD-1 receptor, 184–7
penicillin, 3, 19, 64, 65, 109
Penn, Sean, 190
peptides, 22
perceptual blindness, 34–5
Perry, Katy, 190
Pfizer (pharmaceutical company), 182–3
phagocytes, 40, 42
Phipps, James, 12
pneumonia, 3, 105, 130
Porter, Keith, 36
Powrie, Fiona, 156, 161

prebiotics, 169
probiotics, 169
prostaglandin E2, 114
proteins, 16, 47–9, 70, 73–4, 77, 104, 129, 174, 176, 184–5
  cancer-cell proteins, 53
  co-stimulatory proteins, 48–9, 55, 177, 188
  cytokines, 73, 77–8, 80–82, 84–7, 89, 92–100, 101, 105, 114–15, 132, 161, 175
  receptor proteins, 25, 113, 157, 176–83, 184–6
  stress-inducible proteins, 110–12, 174
psoriasis, 99

radiation, 159–60
Ramon, Gaston, 13
Ramsdell, Fred, 161
receptor proteins, 25, 113, 157, 176–83, 184–6
red blood cells, 38, 43, 59–61, 134
Red Hot Chili Peppers, 190
Reddy, Akhilesh, 146–7
regulatory T cells, 159, 161, 163, 165, 171, 189
Rees, Martin, 29
Reichstein, Tadeusz, 116, 118
Remicade (drug), 98
Revlimid (drug), 195
rheumatoid arthritis, 87–9, 95–7, 99–100, 103, 114, 116–17, 119, 132, 175
Rift Valley fever virus, 58
rituximab, 103–5
Roche (pharmaceutical company), 99
Rockefeller, Laurence, 68
Rockefeller University, New York, 33, 36–7, 52, 93
Roenneberg, Till, 131
Rosenberg, Steven, 78–82, 175, 192
Royal Society, 11, 12–13, 64
Rudensky, Alexander, 161
Russell, Bill, 159–60
Russia, 168

St George's Hospital, London, 12, 165

Sakaguchi, Shimon, 151–8, 161, 169, 189

Sakakura, Teruyo, 151–2

Schering-Plough (pharmaceutical company), 71, 156

Schuler, Gerold, 41–2

Selye, Hans, 121–2

Sendai virus, 67

senescent cells, 141, 142

sepsis, 93–4, 96, 98, 100

Sharma, Padmanee, 172

Shaw, George Bernard, 18

Shaw Prize, 2011, 26

Shevach, Ethan, 158

Simons, Daniel, 34

sleep, 3, 4, 122, 129, 131–2, 134

smallpox virus, 11–13, 29, 58

Social Network, The, (film) 190

space exploration, 134–7, 147

Stanford University, 144

Starling, Ernest, 116

statins, 137

Steamboat Springs, Colorado, 14

Steinman, Claudia, 37, 54

Steinman, Ralph, 26, 32–41, 45, 46–7, 50–56, 177

steroids, 119

Strachan, David, 165

stress, 115, 118, 121–4, 127, 140

stress-inducible proteins, 110–12, 174

Strominger, Jack, 109

sudden cardiac death, 132

suppressor T cells, 154–9

Swigert, John, 135

Szent-Györgyi, Albert, 36

T cells, 16–17, 29, 33, 41, 43–5, 48–9, 51, 102, 105, 109, 130, 143–4, 152, 154–9, 161, 163–5, 170, 176–9, 184, 193–4
  regulatory T cells, 159, 161, 163, 165, 171, 189
  suppressor T cells, 154–9

t'ai chi, 3, 124–7

Taniguchi, Tadatsugu 'Tada', 72

Taylor, Linda, 80–81

telomerase, 140

telomeres, 140

temperature, 112–15

tetherin, 74

TGN1412, 105

thalidomide, 194–5

Thornton, Angela, 158

thymus, 143–4, 152–4, 169

Time magazine, 68

time of day, 130–33

TNF (tumour necrosis factor alpha), 89–90, 170–71
  anti-TNF therapy, 92–106

toll genes, 22–4, 27

toll-like receptor (TLR) genes, 23–5, 27–8, 30, 42
  TLR4 gene, 24–5, 27–8

toll-like receptors, 27–8, 42, 30, 93, 138, 145

toothache, 132

trampolines, 125–6

tuberculosis, 50–51, 101

tumour necrosis factor alpha *see* TNF

ulcerative colitis, 78, 99

University of Basel, 116

University of California, San Diego, 20

University of Chicago, 178

University of Massachusetts Medical School, 126

University of Pennsylvania, 192, 193

University of Texas Southwestern Medical Center, Dallas, 24

University of Zurich, 70

vaccination, 10–16, 29–30, 50–51, 137–8, 145–7
  tailored vaccines, 145

Vilček, Jan, 90–9, 96–100, 178

Vivier, Eric, 186

van Voorhis, Wesley, 41

Wall Street Journal, 191
Walter and Eliza Hall Institute,
    Australia, 83
Walunas, Teresa, 179
Wax, Ruby, 126
Weissmann, Charles, 70–71
white blood cells, 16, 28, 39, 67,
    70–71, 77–8, 177
  see also B cells, T cells, Natural
    Killer cells
Williams, Richard, 94

Woody, James 'Jim', 95–6
Woolf, Virginia, 113
World Health Organization, 58, 182

X-rays, 29

Yale University, 9–10, 20, 23, 153
yellow fever virus, 58, 120
Yervoy (drug), 183–4

Ziskin, Laura, 190